Paul L. Younger has a diverse academic background in applied geosciences and environmental engineering. He currently holds the Rankine Chair of Engineering at the University of Glasgow, Scotland, where he is Professor of Energy Engineering. A Fellow of the Royal Academy of Engineering, he has advised the UK and Scottish governments on various energy issues. His present research focuses on geothermal, hydropower and the environmental management of unconventional gas developments. Paul is a prolific author whose previous books include *Water: All That Matters* (2012) in this series.

ENERGY
ALL THAT MATTERS

ENERGY

Paul L. Younger

First published in Great Britain in 2014 by Hodder & Stoughton. An Hachette UK company.

First published in US in 2014 by The McGraw-Hill Companies, Inc.

This edition published 2014.

British Library Cataloguing in Publication Data: a catalogue record for this title is available from the British Library.

Library of Congress Catalog Card Number: on file.

Paperback ISBN 978 1 473 60188 8

eBook ISBN 978 1 473 60190 1

10 9 8 7 6 5 4 3 2 1

The publisher has used its best endeavours to ensure that any website addresses referred to in this book are correct and active at the time of going to press. However, the publisher and the author have no responsibility for the websites and can make no guarantee that a site will remain live or that the content will remain relevant, decent or appropriate.

The publisher has made every effort to mark as such all words which it believes to be trademarks. The publisher should also like to make it clear that the presence of a word in the book, whether marked or unmarked, in no way affects its legal status as a trademark.

Every reasonable effort has been made by the publisher to trace the copyright holders of material in this book. Any errors or omissions should be notified in writing to the publisher, who will endeavour to rectify the situation for any reprints and future editions.

Typeset by Cenveo® Publisher Services

Printed and bound in Great Britain

John Murray Learning policy is to
recyclable products and made fror
and manufacturing processes are
regulations of the country of origin

John Murray Learning

338 Euston Road

London NW1 3BH

www.hodder.co.uk

Also available
in ebook

To my three sons, Thomas, Callum and Dominic, whose generation will have to cope with the consequences of their forebears' chronic short-sightedness and woolly thinking on energy use.

Contents

Foreword   x

1 Energy: all that matters?   1

2 No, really: energy *is* matter!   11

3 The providence of Prometheus:
  energy resources   23

4 The energy system: securing
  affordable and sustainable supplies   39

5 Upstream activities: preparing for
  energy conversion   53

6 Energy conversion technologies   67

7 Power to the people: energy grids in
  transition   99

8 Secure, green and cheap energy:
  can we really have it all?   113

  100 ideas   125

  Sources and further reading   143

  Index   147

Foreword

Although I am a Professor of Energy Engineering, I would be the first to deny that I am an 'energy expert'. In my view, there's no such thing: the topic of energy is so multifaceted that no one person can truly command the entire subject. Thus, although I have tried my best to cover the principal highpoints of energy in this slim volume, the reader can rest assured that I am the true originator of very little that I present – apart from my obvious personal take on it all.

Source references are listed for all substantive points made. Nevertheless, as with everything I have ever written, the understanding I have gained by osmosis can never be fully acknowledged. Yet try I must, so especial thanks to Dermot Roddy, Michael Feliks, Rob Westaway, Zhibin Yu, Harry Bradbury, Manosh Paul, Ian Fells, Colin McInnes, Tony Roskilly and David Banks. As with all my obsessive writing ventures, the true heroes of the piece are my family, who have to put up with my many hours of mental absence. At times, I have felt like a migratory bird making my way through a giant wind farm. I hope that the end result is not too feather-brained. Either way, my gratitude to family and mentors is genuine and boundless.

Paul L. Younger
Glasgow, Summer 2014

1

Energy: all that matters?

'And for all this, nature
is never spent;

There lives the dearest freshness
deep down things ...'

Gerard Manley Hopkins

▶ Confronting the energy 'trilemma'

Energy is probably *the* defining topic of our age. For many millennia, humankind needed only to ensure access to sufficient affordable energy for its domestic needs: no heed needed to be taken of macro-economics, geopolitics or the atmospheric science of energy consumption. Those days are gone – we urgently need to consider all of these aspects and more. Nonetheless, we have yet to truly grapple with the challenges of energy in a holistic manner. Indeed, for the most part, people in the affluent Global North have grown used to not even having to concern themselves with ensuring access to energy: just flick a switch, turn on the hob or fill the tank and the energy is there, at your service.

This ready access has bred some unhealthy habits, with many people thoughtlessly wasting energy, heedless of the consequences. Those consequences are predominantly borne by others: the less fortunate people who toil to release energy from earth or atmosphere 24/7 – usually with little thanks, and increasingly treated with outright hostility; those as yet unborn whose room for manoeuvre looks likely to be severely straitened by the overconsumption of irreplaceable resources by us, their profligate ancestors; and those non-human beings in our shared ecosystems whose fates are increasingly determined by the polluting and climate-shifting consequences of wanton energy use. There is a clear scientific consensus that the unabated atmospheric emissions of carbon dioxide associated with traditional forms of energy use are (at the very least) severely exacerbating natural variations in climate in undesirable ways.

Ethically, then, the days of heedless extravagance and nonchalance really should be well and truly over. Even if you are sunk in selfishness, there is still an imperative for change, as uncertainty over the long-term availability of some hydrocarbons and nuclear fuels results in price volatility in the world markets. Few credible commentators argue that 'do nothing' is a serious option any longer in the world of energy management. It is far less clear, however, what the 'do something' options ought to include – or what they should definitely *exclude*.

The technical challenges are legion, and annoyingly complicated. Much as it might suit the politicians and campaigners for energy issues to be so simple as to lend themselves to pithy sloganizing, reality is seldom so obliging. Even in the rare cases in which the technical issues are clear cut, the interface with socio-economic constraints is usually riddled with nuances, or even downright contradictions. The conundrums that we currently face have lately come to be known as the 'energy trilemma': how can we simultaneously ensure that energy supplies are reliable, affordable and environmentally sustainable? Satisfying any two of these three requirements may be technically viable, but simultaneously meeting all three is a very tall order indeed. We are unlikely to make much headway simply by wishing things were otherwise; yet wishful thinking has become the default mode in the short-sighted world of energy politics. Sensible, democratic decisions would need to be based on a hard-headed (but soft-hearted) examination of the array of energy options that delimit our freedom to choose.

This book aims to provide the interested, non-specialist reader with just such an examination. In attempting this, it is rather unusual. There are plenty of good books on specific aspects of the energy problem: books on climate change, books on energy technologies, books on energy geopolitics. Yet there is a dearth of authoritative works examining energy holistically, with sufficient coverage of the scientific, engineering, environmental, social and economic dimensions to give a general reader a useful overview – and none at all, until now, in a single, reasonably sized and easily readable volume. If you feel the need for such a book, this is it.

▶ Energy: that's life

Having said that this book is for non-specialists, and having denied that there really is any such thing as an 'energy expert', it is perhaps worth pausing to acknowledge that we are *all* familiar with energy as a fact of life. We have all experienced times when we are bursting with energy, raring to join battle with life's challenges. Similarly, we have all experienced the listlessness born of enervating illness. Our bodily experience of energy is very real: as mammals, our bodies depend utterly on the efficient conversion of fuel (i.e. food) into muscular and nervous energy. Illness is any condition that upsets this process. The same applies to all living creatures: the impulse to harness, convert and expend energy is ubiquitous on our wonderful living planet.

So intimate has the human relationship with harnessing edible energy become that, according to Harvard primatologist Richard Wrangham, the very evolution of higher cognitive powers in *Homo sapiens* appears to have been heavily influenced by the emergence of cookery: the ability to concentrate nutrients and render them more readily digestible left our ancestors with far more time for thinking and other creative activities than our closest relatives among the primates, who spend a large proportion of their waking hours finding and eating raw food, with its far lower energy yield per mouthful. So it turns out that the use of energy for cooking is no longer an option for us. We have evolved beyond a point of no return.

We frequently hear the equation 'water is life': we could just as reasonably claim that energy is life. The most archaic life forms still present on our planet, the obligate lithoautotrophic microbes, make their living by diverting for their own use some of the energy that naturally circulates between oxidized and reduced forms of iron, sulphur and other elements. So accomplished have they become at this diversion that they have devised mechanisms that accelerate these otherwise inorganic processes by as much as a million times. Higher plants have devised amazingly efficient means of intercepting solar radiation and using this to combine atmospheric carbon dioxide and soil-sourced nutrients to produce cellulose, lignin and other tissue-building compounds.

▶ Energetic expressions

Given the ubiquity of energy in all of our lives, it is unsurprising that the vocabulary of energy has gradually diffused throughout the lexicon of modern society. From the proclamation of the intellectual maestro of the US Founding Fathers, Benjamin Franklin, that 'Energy and persistence conquer all things', to Homer Simpson affirming, in response to Lisa having claimed to invent a perpetual-motion machine, that 'in this house we obey the laws of thermodynamics', references to energy are everywhere. A convincing case has been made that the publicly acclaimed advances in physics in the nineteenth century caught the imagination of generations of poets and novelists, amounting to a veritable 'thermopoetics', as Barri J. Gold has termed it. Thus, for instance, Gerard Manley Hopkins (1844–99) in his poem 'God's Grandeur' wrote:

> *The world is charged with the grandeur of God.*
> *It will flame out, like shining from shook foil;*
> *It gathers to a greatness, like the ooze of oil …*
> *… And for all this, nature is never spent;*
> *There lives the dearest freshness deep down*
> *   things …*

These words, and many more of Hopkins' poems, crackle with exuberance as they bear witness to the flows of energy in Nature. Consciously or not, poets such as Hopkins gave elegant voice to the world-view of the pioneer thermodynamicists of his age, such as Lord Kelvin, for whom (as Marie Hadfield has noted):

> *the laws of nature were instruments of the divine will, which alone had power to alter or destroy ... 'The energies of Nature' were for [Kelvin] the Creator's gifts, which offered to humanity opportunities to direct the universe ... In his argument, the idea of God as architect or designer, as in Natural Theology, is replaced by the idea of God as the source of all power in nature ...*

Less comely than the energies of Nature were the noise and dirt of industrial energy use, through which, Hopkins went on,

> *... all is seared with trade; bleared, smeared with toil;*
> *And wears man's smudge and shares man's smell ...*

The industries of energy – mining and mechanical engineering – provide dystopian scenery and vivid metaphors for the alienation of colliers and others involved in such trades:

> *Within his head revolved a little world*
> *Where wheels, confusing music, confused doubts,*
> *Rolled down all images into the pits ...*
> *Within his head the engines made their hell,*
> *The veins at either temple whipped him mad,*
> *And, mad, he called his curses upon God ...*

> *Where, what's my God amongst this crazy*
>   *rattling ...*
> *Man to engine, battling, bruising,*
> *Where's God's my Shepherd, God is Love?*
>
> Dylan Thomas, 'Out of the Pit'

As we shall see, notions of an Eden lost through fossil-fuelled industrialization and hankerings after a 'return to nature' are powerful currents in the early twenty-first-century energy debate.

▶ About this book

Before diving into the emotive debates over energy futures, this book takes a dispassionate look at the science and engineering of energy. First up is a hugely simplified introduction to the physics of energy (Chapter 2), exploring how our everyday experiences of energy mesh with the initially counterintuitive finding that energy and matter are ultimately interchangeable. Having paid due homage at the shrines of Newton, Einstein, Clausius, Kelvin and others, the remainder of the book focuses on *engineering* rather than pure science. Thus, we first quantify those energy resources available on our planet – renewable and otherwise – (Chapter 3), then proceed to consider what needs to be done to take advantage of them (Chapter 4), before explaining how this is actually achieved (Chapters 5 and 6). We then consider how electrical power is delivered in modern, predominantly urban, societies (Chapter 7), which in large measure determines our room for manoeuvre when imagining different ways of converting and delivering energy in future. Chapter 8 places all of the above in the context of the energy controversies of today, asking whether we can really have energy supplies that are secure, green *and* affordable. As with all books in this series, this volume ends with a section entitled '100 ideas', which is a resource base of information and links to allow you to further explore the key issues introduced in this brief presentation.

2

No, really: energy *is* matter!

'... Thermodynamics is a funny subject. The first time you go through it, you don't understand it at all. The second time you go through it, you think you understand it, except for one or two small points. The third time you go through it, you know you don't understand it, but by that time you are so used to it, it doesn't bother you anymore.'

Arnold Sommerfeld (1868–1951)

▶ What can the matter be?

$E = mc^2$

The most famous equation in the world is surely Einstein's deceptively simple expression

$$E = mc^2$$

This originally arose from his exploration of the consequences of his much discussed theory of special relativity. Unpacking this equation, it means that energy (E) is equal to mass (m) times the 'square' of the speed of light (c). Let's try to be clear on what is meant by each of these three things.

First, let's be clear what we mean by energy (E): **energy is the ability to perform work.** In other words, it is a property of some object (or 'physical system') that enables it to change the status of some other object(s) (or 'physical system(s)'). To give an obvious example, a hammer of a given mass, moving at a given speed, has the energy necessary to drive a nail a certain distance into a plank of wood. We could, if we had the time and inclination, calculate just how much energy was transferred by the hammer to the nail, and then onwards into the splitting of the wood, in order to achieve an observed incremental advancement of the nail into the wood. If we had even more time, we could account for the energy used by the person wielding the hammer, and trace that back through their foodstuffs to photosynthesis and thus, ultimately, to the Sun. In doing so, we would have traced energy through a whole series of forms, from kinetic energy (in the hammer) through chemical energy (in the body) to radiant energy (captured by plants) to nuclear energy (in the Sun). This isn't even the full list of energy forms in this work-achieving chain of events. But it underlines the point that 'ability to perform work' comes in multiple guises, and that energy is not 'created' anywhere along the chain, but merely converted from one form to another.

The forms into which energy is converted *en route* are not all useful. For instance, the noise (acoustic energy) produced by striking the nail with a hammer is of limited use: while the tone of the noise might tell the hammer-wielder whether they have hit the nail rather than the wood, to a neighbour it is likely to come across as nothing but a nuisance. Similarly, the nail is likely to heat up a little when struck by the hammer, and the wood will also heat a little as the nail prises it apart. Thus, the total amount of energy expended by the hammer-wielder is more than is manifest in the useful work of the nail entering the wood.

As energy occurs in so many forms, so it has been quantified using a bewildering variety of units; to this day, the technical community is somewhat resistant to standardization. As long ago as 1967, the eminent physicist Richard P. Feynman (1918–88) quipped: 'For those who want some proof that physicists are human, the proof is in the idiocy of all the different units which they use for measuring energy.' We will meet many of these different units later in the book. For the time being, we will just note the most fundamental of them all – the joule – one unit of which is just about the amount of energy I expend lifting my smartphone from my pocket to my ear. (The formal definition of a joule is rather more complicated; see box.)

Two takes on a joule

Named after a pioneering English physicist, James Prescott Joule (1818–89), the SI unit for energy is commonly defined in two rather neat ways:

For the mechanically minded: One joule is the amount of energy that it takes to exert a force of one newton over a distance of one metre.

For the electrical engineer: One joule is the energy required to pass an electric current of one amp(ere) through a wire with a resistance of one ohm for a time period of one second.

What about mass (m)? We are all personally familiar with this phenomenon in daily life. As too many members of the sedentary, urban populations of the world ingest more energy as food than they expend in exercise, they experience an increase in body mass. In other words, they 'put on weight' or 'get fat'. In common parlance, 'mass' is used interchangeably with 'weight', though physicists draw an important distinction between the two: **mass is the amount of matter that something contains**, whereas the weight of something of a given mass is the magnitude of the gravitational force that it experiences. As you would expect, the two properties have different units. For instance, a bag of sugar with a mass of 1 kilogram will actually have a 'weight' (in formal terms) of 9.81 newtons – a 'newton' being the internationally agreed unit of force, which is equivalent to 1 kilogram subject to the average acceleration due to gravity on planet Earth of 9.81 metres per second squared (usually written m·s^{-2}). Here's the interesting bit: the same object would have exactly the same *mass* were it placed on the surface of the Moon, but since the gravitational attraction of the Moon is only about a sixth of that of the Earth, the *weight* of the object would only be 1.635 newtons. This is, of course, precisely why astronauts float around inside the International Space Station. So, next time you 'weigh' anything in the kitchen, bear in mind that the read-out is actually in units of mass, not weight. In the rest of this book, the term 'mass' will always be used instead of 'weight', but, in line with common parlance, if we talk about 'weighing' something we are really talking about measuring its mass.

So where does all this leave us in defining mass? Well, the use of the word 'newton' for the unit of force is no coincidence, but reflects the fact that Sir Isaac Newton's Second Law states that force (F) is equal to mass (*m*) times acceleration (a), that is:

$$F = m\,a$$

This is almost as famous an equation as $E = m\,c^2$. It allows us to propose an alternative, more formal, definition of mass as follows: **mass is the resistance a given object offers to acceleration by an external force of a specified magnitude.** This resistance is closely related to its atomic properties – that is, how many neutrons, protons and (less importantly) electrons each atom of that object contains – which brings us back to the notion that mass equates to how much matter an object contains.

Two down, one to go: c stands for the **speed of light** as measured in a vacuum, which is a very large number: 299,792,458 metres per second. If we remember our primary school maths, c^2 means that this value is squared – that is, multiplied by itself. The result is truly humungous: 89,875,517,873,681,800. So energy equals mass times this huge number, representing the square of the speed of light in a vacuum. Fine, but why should the speed of light come into it? Why should the relationship between energy and matter be somehow mediated by the behaviour of light? The reality is that light in and of itself doesn't really have much to do with it; it is simply a convenient 'yardstick' that quantifies a fundamental property of the universe: nothing can go faster than light travelling in a vacuum. This is because photons, the entities that constitute light, have a full complement of energy but no mass. This means that light can whizz around like a reckless driver who always keeps their foot on the pedal. As Professors Brian Cox and Jeff Forshaw succinctly put it:

> *The apparent specialness of light is an artefact of our human tendency to think of space and time as different things ... everything hurtles through spacetime at the same speed, c, including you, planet Earth, the Sun and the distant galaxies. Light just happens to use up its spacetime speed quota on motion through space, and in so doing travels at the cosmic speed limit.*

So the speed of light is simply a proxy for the cosmic speed limit in the continuum of 'spacetime', and this speed limit turns out to exert a fundamental control on how energy and matter interrelate.

Going back to Einstein's famous equation, then, what does this all mean? In essence, the expression **$E = m c^2$ is telling us that an object with a very small mass actually embodies an enormous amount of energy.** Let's use the equation and see how much. If we stick to joules, kilograms, metres and seconds, the units should all be mutually consistent. So, if I have a walnut, say, with a mass of 15 grams (i.e. 0.015 kg), then it contains:

$$0.015 \times 89{,}875{,}517{,}873{,}681{,}800$$
$$= 1{,}348{,}132{,}768{,}105{,}230 \text{ joules of energy}$$

This is an enormous amount of energy. It is so large that we are better using aggregated units of joules to express it. You may recall from other fields that a thousand (10^3) is *kilo-*, a million (10^6) is *mega-*, a billion (10^9) is *giga-*, a thousand billion (10^{12}) is *tera-*, and a million billion (10^{15}) is *peta-*. So this amount of energy is 1.348 petajoules. To give a flavour of its magnitude, this is a tad more than the average monthly electricity use in the entire United States. So why don't we just extract vast amounts of energy from walnuts to meet our energy needs?

To find out, let's do a little experiment: we'll put the walnut in a special oven fitted with devices that can measure energy output (actually just glorified thermometers), burn it away to nothing, and see how much energy it yields. Brace yourself for a disappointing answer: 180 kilojoules, or just 1.8×10^{-10} petajoules. This is about 10 per cent of the total energy stored in an everyday AA battery, but only about a billionth of a per cent of the energy that Einstein's equation tells us was actually stored in that walnut.

So was Einstein wrong? No. If you could arrange a further experiment, somehow adding 90 terajoules of energy to our walnut without setting it alight, it would actually gain a gram in mass, now weighing in at a full 16 grams. So mass and energy are indeed interchangeable. (This insight is unlikely to change the fortunes of nut farmers anytime soon, however, as 90 terajoules is roughly equivalent to a full load of fuel in a jumbo jet!) But why did we get so little energy from our walnut when we burned it away to nothing? The answer is: we didn't really burn it away to *nothing*, we vaporized it: we broke a large number of chemical bonds *between* atoms, releasing 180 kilojoules in the process, and dispatching a lot of atoms that made up the solid walnut in the form of gases, mainly carbon dioxide and water vapour. This made the walnut disappear to human eyes, but actually left intact the internal energy *within* the atoms of carbon, hydrogen and oxygen. This is far and away the bulk of the energy that is predicted to be present using the equation $E = mc^2$.

Can we not find a better way than burning the walnut that will let us get at the energy stored in these atoms? Sadly not: there is no commonplace, everyday way to harness the energy that Einstein's equation correctly tells us is embodied within all matter. There is one notable exception: nuclear reactions. This is of overwhelming importance to us, as it is natural nuclear reactions that produce all of the energy we receive from the Sun, and which account for the bulk of the geothermal energy that spontaneously arises within the Earth's interior. Since the 1940s, we have also been able to harness nuclear reactions to artificially produce energy. At first, the creative genius that unlocked this capability was applied solely for destructive purposes, in the form of nuclear weapons. But harnessed in civil nuclear power plants, exploitation of the truth embodied in the equation $E = mc^2$ has produced vast amounts of energy from very modest amounts of enriched uranium – and with negligible atmospheric carbon emissions. This is something we will return to in Chapters 3 and 5. However, most materials are not prone to nuclear fission in the way that uranium is, and we simply can't get at more than a tiny fraction of their embodied energy – at least not yet, and perhaps not ever.

▶ No free lunch: the laws of thermodynamics

By now, after decades of adulation for Einstein and a steady stream of popular TV programmes on astrophysics, most people have at least heard of the laws of thermodynamics. Indeed, as we've already noted, even Homer Simpson professed familiarity with them, way back in 1995. Yet the name 'thermodynamics' is rather off-putting to many, suggesting something intensely technical and obscure. This inhibits many people from even trying to understand these laws. Yet, in essence, the concepts are rather simple and straightforward. It doesn't help that there is lack of consistency even in the statement of how many laws of thermodynamics there are. Most physicists recognize four, but confusingly the first and second laws come second and third respectively in the full list. (The first one in the list was added later and is known as the 'Zeroth' Law, i.e. law number zero.) For our purposes, only two of the laws are of crucial importance: the First and Second laws.

The **First Law of Thermodynamics** can be stated in many ways, but perhaps the most useful formulation is the following: **energy cannot be created or destroyed – it can only change (or be changed) from one form to another.** This has the immediate practical application that the energy budget of a given process will always balance. Going back to our walnut, we saw that the total amount of energy it contained before we completely burned it was 1.348 petajoules. The process of burning released 1.8×10^{-10} petajoules in the form of heat. This means that 1.34799999982 petajoules of energy remains trapped in the carbon dioxide and water vapours that were released during burning. These sorts of calculations are invaluable in all fields of energy engineering, as they allow us to predict the energy yield from fuels (or, say, wind) and account for energy that is 'lost' to us, in that it is converted into an unusable form.

The notion of conversion of energy to a less useful form underpins the **Second Law of Thermodynamics**, originally identified by Rudolf Clausius (1822–88), which can be summarized as follows: **there is an overall tendency for energy to be converted from high-quality forms to low-quality forms.** In practical terms, this means, for example, that there is a tendency for flow of heat to occur from hot areas towards cold areas. So your coffee cools over time as energy is transferred from the artificially concentrated high-temperature environment of your cup into the unruly realms of the cooler air. This is a rather obvious example, but in the case of chemical reactions the result may not be so intuitive. Hence, knowing that there is a preferred direction for energy transfer is extremely useful in many engineering analyses. In the formal terms of physics, this directionality is expressed as follows: the entropy in the universe is gradually increasing. '**Entropy**' is not an everyday term, but it has much in common with more familiar terms such as 'disorder' and 'scattering at random'. Essentially, the cosmic implications of the Second Law of Thermodynamics are that the universe will eventually die a cold death, when all the higher-quality (or 'ordered' or 'structured') forms of energy (and therefore matter) are evenly distributed throughout the cosmos, at a temperature of a few degrees above absolute zero. (This incidentally, is a loose statement of the Third Law of Thermodynamics.)

The implications of all of this are that, in energy budgeting, you cannot get anything for nothing, and even the most efficient energy conversion process will result in some of the energy being 'lost' into unusable forms. This is highly reminiscent of the common aphorism that 'there is no such thing as a free lunch', and many people use this expression as a colloquial expression of the Second Law. When we consider biological systems, at first glance it can appear that low-quality forms of energy (e.g. heat) turn into higher-quality forms (e.g. chemical energy in plant tissues). However, all of this entails unseen wastage of energy in the process, and the net effect over millennia is still as the Second Law tells us.

It is worth noting in passing that this simple summary has merely skimmed the surface of the vast domain of thermodynamics, and, in doing so, has not even introduced some of the key considerations that must be included in any quantitative analysis. For instance, in talking of energy flows, we have spoken of heat, but we have not emphasized the crucial importance of specifying the pressures and volumes within which the hot substances are constrained. When these factors are properly included, we begin to deal in a property called **enthalpy**, which is the energy contained in a system plus the product of its pressure and volume. Accurately accounting for changes in enthalpy is crucial when we analyse chemical energy transformations, or the energy yield of naturally hot, pressurized water in deep geothermal reservoirs, to give but two examples. For now, though, we'll simply note the need for caution in calculations and move on to considering what energy resources are available to us here on planet Earth.

3

The providence of Prometheus: energy resources

'... when the other gods were away, [Prometheus] approached the fire of Zeus, and with a small coal from this shut inside a fennel-stalk, he came away joyfully, seeming to fly rather than run ... Up to this time, then, men who bring good news usually come with speed, after the manner of Prometheus ...'

Pseudo-Hyginus, Astronomica *2.15, second century CE*

The legend of Prometheus' theft of fire is but one of many similar stories from the ancient cultures of Eurasia and the Americas that seek to explain how humans came to possess the striking ability to manipulate thermal energy. Significantly, the Prometheus story anticipates the recent scientific conclusion that humans would never have flourished without the benefits of cooking. It also highlights the joy that ready access to energy has brought to countless generations of humans. Finally, it traces the ultimate source of earthly fire back to Zeus's fire – which was identified with the Sun in Greek mythology. This identification is significant, as so many of our energy resources can indeed be traced back to the Sun – and the few that can't still owe their origins to the same astrophysical events that gave rise to the Sun. In this chapter, we'll explore the ultimate origins, nature and magnitude of the energy resources potentially available to humankind – and their carbon footprints. This will make it easier for us subsequently to consider technologies for energy conversion (Chapters 4, 5 and 6) and delivery (Chapters 4 and 7).

▶ Where from?

Energy *sources*

It is possible to ascribe all usable forms of energy on our planet to only three ultimate sources: solar radiation, gravity and radioactivity. Given that the production of solar radiation itself depends on radioactivity, you could reduce the list of sources to just two. Viewed yet another way, as gravity is a function of the size and disposition of the Earth and other planets, which in turn depend critically on the behaviour of the Sun, we could say that all energy sources are ascribable to the presence of the Sun and its satellites. However, both of these reductionist options would drive our narrative down a line of enquiry that would become ever more remote from energy as experienced by human beings, so we'll settle for the three sources: the Sun, gravity and radioactivity. The various energy sources map on to these three, as shown in Figure 3.1.

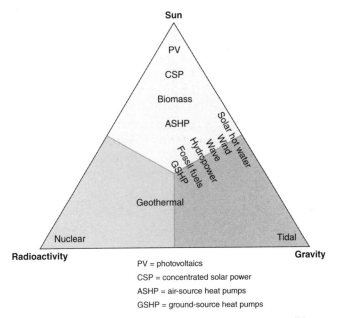

PV = photovoltaics
CSP = concentrated solar power
ASHP = air-source heat pumps
GSHP = ground-source heat pumps

▲ Figure 3.1 Ultimate origins of major energy sources accessible to humankind. All of these energy conversion techniques are explained in Chapter 5. Where the named energy sources overlap more than one field, they draw part of their potency from that secondary source.

Sun

The most obvious source of **energy from the Sun** is solar radiation, which is the form of energy captured directly by photovoltaic panels and solar hot-water systems, and indirectly by concentrated solar-power plants (all discussed in Chapter 6). Solar radiation is indirectly responsible for most of the other well-known renewable energy sources, too, in the following ways:

▶ **Wind:** differential solar heating of different parts of the Earth's atmosphere creates the differences in air pressure and buoyancy that result in winds.

▶ **Wave:** the shearing action of winds on the water surfaces of oceans and lakes gives rise to waves, so to the extent that winds are driven by solar radiation, so are waves.

▶ **Hydropower:** evaporation of water is driven by solar radiation, and because this is the indispensable precursor of the precipitation that hydropower ultimately relies upon, the Sun is also the ultimate source of hydrological resources.

▶ **Air-source heat:** solar warming of air provides the heat that is ultimately recovered using heat pumps (see Chapter 6).

▶ **Ground-source heat:** this resource is principally supported by solar warming of rainwater that then infiltrates into the soil and underlying rocks, which is, again, harvested using heat pumps.

▶ **Biomass:** plants grow by means of photosynthesis, which involves capturing solar radiation to drive the biochemical reactions that combine carbon, water and other nutrients to produce biomass, which can ultimately be burned or gasified for energy use.

It is less common to think of **fossil fuels** as solar in origin, yet the vast majority of fossil fuel deposits owe their ultimate origins to biomass that accumulated many millions of years ago.

Gravity

Gravity is responsible for the distortion of the depth of seawater, which results in the diurnal tides encountered all around the world's ocean shores. Three heavenly bodies play a role in this:

▶ **The Earth's** own gravitational field draws the ocean water towards it, thus preventing it being swept away into space.

▶ **The Moon** exerts the bulk of the external gravitational tug that piles the oceanic waters up towards the side of the Earth nearest to it.

▶ **The Sun's** gravitational attraction modifies the degree of distortion of water distribution wrought by the Moon: when the two are aligned, the distortion is maximized and spring tides are experienced; conversely, when the two are at polar opposites, the Sun reduces the effect of the lunar attraction and neap tides result. This is why it would be wrong to claim that solar radiation is the sole source of energy from the Sun – solar gravitational force is also important.

Clearly, tidal energy is totally dependent on the interplay of these three gravitational fields.

The Earth's gravity field also plays an important role in mediating several other energy sources, most notably as follows:

▶ **Hydropower:** gravity not only leads to water droplets falling as rain; it is also crucial in finally driving flowing water through the turbines that extract power from it.

▶ **Wind:** gravity interacts with heat to move air and produce winds.

▶ **Wave:** heaving and pitching motions in waves develop substantially in response to gravity.

▶ **Ground-source heat:** gravity is responsible for driving air-warmed waters into the soil and on downwards.

▶ **Geothermal:** similarly, gravity-driven movement of naturally infiltrating waters can also be important in geothermal systems, and is certainly important wherever geothermal fluids are reinjected as part of the energy conversion cycle (Chapter 6).

▶ **Solar hot water:** many roof-mounted systems in warm regions operate by a combination of thermal convection and gravity alone (Chapter 6), without any need for an auxiliary pump.

▶ **Fossil fuels:** gravity played a crucial role in the accumulation and compaction of organic debris and enclosing inorganic sediments that comprise coal-bearing strata and hydrocarbon provinces.

Radioactivity

As we've already noted, **radioactivity** in the Sun is the ultimate source of solar radiation; closer to home, it is also the predominant source of heat generated within our planet, which is manifest in deep-sourced geothermal energy. Other sources of energy also contributed to geothermal energy earlier in the Earth's history, such as nebular heat released during the coalescence of Earth in the first place around 4.5 billion years ago, gravitational heating as the Earth expanded, and kinetic energy transmitted from incoming asteroids and meteors, which were far more abundant in the early history of the solar system. However, the bulk of geothermal energy nowadays is radioactive in origin. To the extent that ground-source heat pumps access genuine geothermal energy, rather than sun-warmed infiltrating water, they also can be said to draw energy from radioactive sources.

The radioactive origin of deep subsurface heat *does not* mean that geothermal energy is hazardous; natural radioactive decay is a rather diffuse phenomenon that releases appreciable heat but relatively few potentially troublesome radioactive daughter products. Moreover, the isotopic form of uranium responsible for geothermal heat (i.e. ^{238}U) is entirely different from the isotope used for nuclear-power generation (i.e. ^{235}U).

We saw in Chapter 2 that vast amounts of energy are locked up in the atoms that constitute all tangible matter in our universe. Equally, we saw that very few of these atoms give up their energy readily. It was only with the search for nuclear weapons during World War II that the technology was developed to take advantage of the few promising categories of atoms, by manipulating the rays and particles emitted by decaying radioisotopes to prise apart other atoms, releasing vast quantities of energy in the process. **Nuclear energy** is thus entirely artificial. While there are a few geological deposits around the world that display some signs of having once hosted spontaneous, natural, nuclear fission, these are small, and the periods of fission they supported were clearly very short lived. The reason is that the concentrations of fission-prone atoms (equal to or less than 0.72% ^{235}U) are too modest in even the richest natural deposits to sustain the chain reactions that are essential to the release of usable quantities of energy. Hence, even if these exemplars of natural fission had still been operative today, they would not represent significant energy resources. It is only with the mining and artificial enrichment of uranium that fission can be initiated and maintained at a significant scale.

The degree of enrichment necessary to support nuclear energy production (i.e. about 3.5% ^{235}U) is far less than that needed for weapons production (i.e. equal to or less than 20%, and usually far greater at more than 85% ^{235}U). Thus there is no necessity to link civilian use of nuclear power to military weapons programmes. Nonetheless, this is precisely what was done in those countries that now possess nuclear arms, and it is also why uranium enrichment initiatives in other countries are closely monitored (e.g. the recent controversies over Iran's nuclear ambitions): any enrichment of uranium beyond the 20% ^{235}U threshold clearly implies a desire to develop nuclear weapons.

▶ How much? Energy resources

Energy sources only become *resources* when they can be regarded as quantifiable forms of energy that are reasonably accessible for use by humans. Resource estimates differ from one analysis to another, however, depending on choices made about what constitutes 'use' and 'accessibility'. With regard to use, the mass media (and therefore most politicians and activists) persistently use the term 'energy' as if it were synonymous with 'electricity'. Yet the use of heat and fluid transport fuels worldwide dwarfs electricity consumption. Unless specified otherwise, when the term 'energy' is used in this book, it refers to *all* usable forms of energy: electrical, thermal and transport fuels. It turns out that different energy resources to some degree lend themselves to particular end-uses. For instance, geothermal resources that are not hot enough to generate appreciable electricity can still deliver vast quantities of renewable heat. On the other hand, wind turbines generate solely electricity, and can only contribute to heating via electrically powered heaters, which (with the exception of the most efficient ground-source heat pumps) consume electricity extravagantly and are consequently very expensive to run. Solar radiation can be harnessed by two entirely different technologies to directly produce electricity (the *photovoltaic* effect) or hot water.

What about **accessibility**? The practicality of retrieval of energy from remote locations is a very real constraint, however attractive the energy resources in themselves might otherwise seem: if it requires more energy to tap an energy resource than it will actually yield, then the exercise becomes pointless. For instance, the vast majority of waves occur in open seas thousands of kilometres from land and are thus of little practical use to us, given the massive infrastructure requirements that their harnessing would demand. Similarly, the vast bulk of geothermal heat is present at such great depths that we are never likely to be able to access it, given the temperature and pressure constraints on the use of metal drilling equipment and so on. This is not to say that wave or geothermal resources are negligible, however: both can be found abundantly in far more accessible locations.

For very high-value commodities such as oil, accessibility is less of an obstacle. Hence offshore oilfields have been productive on a large scale since the 1970s, and such technology is gradually spreading into deeper ocean areas and higher latitudes. This, however, has raised a further dimension of accessibility: given other priorities – especially the conservation of fragile ecosystems – there is a strong case for governments to limit accessibility, even where the economics would otherwise favour it. For instance, a world-beating geothermal power plant could be developed at Yellowstone National Park, but not without detracting from the park's wilderness atmosphere, and not without risk of deterioration of some of the natural hydrothermal features on which the park's identity relies.

Having attained clarity on use and accessibility, we next need to quantify energy resources. Ultimately, we need to calculate how much of a given energy resource we need to harness to meet a given energy demand. Such calculations allow us to compare the productivity of alternative energy sources. This can be expressed in a number of ways. For instance, **energy density** expresses the amount of energy per unit volume of fuel. A similar concept, known as **specific energy**, quantifies the energy content per unit mass of a given substance. (In both cases, the energy content usually refers just to the usable energy, not the total energy embodied within the material which would be calculated from $E = mc^2$.) Those two metrics are logical enough for traditional combustible fuels, but they make less sense for many renewables, which do not use volumes or masses of fuel, but simply gather naturally occurring forms of energy passing over a given area of land. For wind and solar energy resources, for instance, it is more apposite to quantify **power density**, which is the energy produced per unit time over a given surface area. Even that measure is tricky to apply in many cases, however. For instance, if farmland is planted with crops destined for fermentation to produce ethanol (as a renewable liquid transport fuel), then clearly that land cannot be used for any other purpose until after harvest. More enduringly, large solar farms, in which significant areas of countryside are covered with photovoltaic panels or parabolic mirrors, can preclude other simultaneous uses of that site for as long as the farm exists. However, where solar panels are installed on roofs of buildings, they do not eliminate other uses of the space beneath. Similarly, land draining to a hydropower dam will certainly be used for other purposes, farming can continue in the shadow of wind turbines, and land underlain by underground coal mines and oil reservoirs continues to be used for all manner of other purposes.

There are formulae we can use to convert power densities into specific energies or energy densities, but these all involve making assumptions about the most appropriate volume to analyse. For instance, although a solar panel has an obvious surface area, it also possesses a certain thickness, by which it can be multiplied to produce a volume. But is this the right volume to analyse, or should we be analysing the entire volume of space between the Sun and the solar panel? A further problem arises from the fact that a volume of space through which solar radiation travels will be reused for the same purpose indefinitely, whereas, once a given volume of hydrocarbon has been burned, it has gone for ever. On the other hand, it is possible to burn hydrocarbons on demand, 24/7, whereas sunlight occurs only during daytime. Thus, however fruitful solar energy might be at noon, its whole-day yield will have to take into account a global average of 50-percent hours of darkness. Converting specific energies or energy densities into power densities is beset with similar complications. These are genuinely difficult to resolve, and the assumptions made in solving them can be tainted by the prejudices of the analyst. However, whichever way you look at it, a certain hierarchy of power density emerges, as illustrated in Figure 3.2.

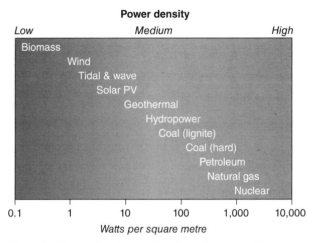

Power density

Low   *Medium*   *High*

Biomass
 Wind
  Tidal & wave
  Solar PV
   Geothermal
    Hydropower
    Coal (lignite)
     Coal (hard)
     Petroleum
      Natural gas
      Nuclear

0.1 1 10 100 1,000 10,000

Watts per square metre

▲ Figure 3.2 Power densities of the principal commercial (or emerging) energy resources. While differing assumptions lead to variance over absolute magnitudes (marked on the bottom axis), there is little dispute over relative rankings.

In broad terms, the lower the power density, the greater the land area you would have to use to obtain a certain power supply. If minimization of land-take were all that mattered, then the work of the energy engineer would be simple. However, power density alone will not suffice. For one thing, the fossil fuels and nuclear will all eventually run out, a prospect that is already revealed in fluctuating prices, as proven geological reserves wax and wane. Furthermore, if fossil fuels are used without abating their associated atmospheric emissions of carbon dioxide, unacceptable climate change will result.

▶ Putting your foot in it: carbon footprints, fossils and renewables

It is rather depressing that, with the sole exception of nuclear, the power density of various energy resources is just about in inverse proportion to their carbon footprints. Here's another summary diagram (Figure 3.3), this time ranking the principal energy resources by carbon emissions. This is an easier comparison to make than

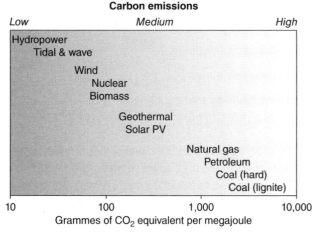

▲ Figure 3.3 Carbon dioxide emissions of the principal commercial (or emerging) energy resources, based on a thorough review of rigorous published studies undertaken by the Intergovernmental Panel on Climate Change (IPPC). (To convert these numbers to the equivalent values per kWh (with which many people are more familiar), simply divide by 3.6.)

power density, as we can in all cases divide the amount of CO_2 produced over the full life cycle of the technology by the amount of energy produced.

The principal reason why the current debate around energy futures is so polarized is precisely that some people regard the above ranking of energy resources by carbon emissions as a non-negotiable priority list for resource selection, whereas others argue on technical and financial grounds that the earlier ranking by power density must take precedence. Who is right? Or is some compromise position the best outcome? Even if it were, is such a compromise achievable? We will spend much of the rest of the book examining crucial details that must inform any rational judgement on this topic, before finally addressing these questions directly in Chapter 8.

In the meantime, it is important to note that the carbon emissions of the fossil fuels are potentially amenable to significant abatement by means of **carbon capture and storage (CCS)**. The principles of CCS are simple enough: CO_2 is extracted from the exhaust gases of a power station and injected by means of boreholes into deep underground strata, which have the capacity to retain it for more than a thousand years. In many cases, retention of CO_2 is expected to be aided by geochemical reactions that lead to the formation of solid, immobile, carbonate minerals. Although the component technologies of CCS are all well established, suitable strata are not ubiquitous. This at least implies that CO_2 may need to be transported long distances by pipeline (for which, again, the technology is well established) prior to injection. Where no suitable strata can be accessed within a reasonable pipeline distance, alternatives to CCS might conceivably include **productization**: using the CO_2 as a chemical feedstock to form stable solids with long lives – such as many plastics and materials used in building foundations. However, large-scale implementation of these approaches requires the attainment of sufficient political will to overcome the cost barriers. As we shall see in Chapter 8, these barriers are not insurmountable; the consensus is that they would simply put the cost of fossil fuels on a par with mainstream renewables such as onshore wind.

4

The energy system: securing affordable and sustainable supplies

'What is a cynic? A man who knows the price of everything and the value of nothing. And a sentimentalist is a man who sees an absurd value in everything and doesn't know the market place of any single thing.'

Oscar Wilde

Energy discussions nowadays are extraordinarily prone to reductionism. If it is not the reduction of 'energy' to electricity, then it is the reduction of all environmental issues to carbon emissions; or if it is not the reduction of energy resource evaluation to gross figures on total solar radiation alone, then it is the reduction of supply – demand balance to domestic insulation. In the preceding sentence alone, I have probably managed to alienate at least four groups of readers. Let me be clear: of course, there *are* strong grounds for advocating all of these things; it is just that achievement of any lasting improvements in the way energy is used will demand that we resist the temptation to seek a simple, single panacea. No one technology or social arrangement will deliver; we simply *must* consider energy as a complete *system.* When we use the word 'system' in this context, we mean a single entity comprising multiple, intimately interconnected elements. In a system, it is usually impossible to make changes to any one element without having knock-on effects on (an)other element(s), and therefore on the behaviour of the system as a whole. In this chapter, we will briefly consider what constitutes the energy system, paving the way for more detailed discussion of its key elements in succeeding chapters.

▶ Elements of the energy system

In broad terms, the energy system can be conceived as having two subdivisions, which are usually known as 'the supply side' and 'the demand side'. The supply side includes all of the technologies of energy conversion, storage, transmission and distribution, whereas the demand side is concerned more with socio-economic phenomena that govern how much energy is desired for specific purposes. Figure 4.1 summarizes the overall

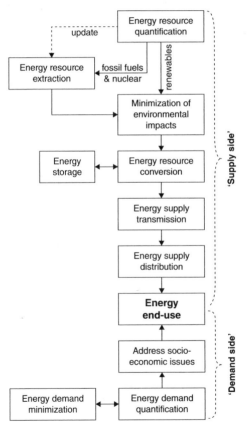

▲ Figure 4.1 The energy system – a simplified summary.

energy system, further breaking down some of the component elements of the supply and demand sides. As the diagram shows, energy end-use is where the two sides come together, and where they must interface in a rather complex socio-technical manner. In round terms, the overall goal of energy engineering is to try to ensure that supply and demand are comfortably in balance at the point of end-use. Although this seems simple enough in principle, the detail is fraught with complex challenges, which are typically addressed by compromises and trade-offs. It is the distaste that absolutists feel for the very notion of compromise that gives rise to much of the heat in the energy debate. Yet, without significant compromises, the system will grind to a halt.

▶ Balancing demand and supply: isn't storage the answer?

Storage is the typical solution to imbalances of supply and demand in most other spheres of life. For instance, winter rains are stored in reservoirs to ensure water supplies through summer, and the late-summer harvest is stored in granaries to feed beasts and humans through the barren winter. So why should energy be any different? Balancing demand with supply would indeed be straightforward were **energy storage** always simple and cheap to arrange.

In some cases, storage of energy is feasible, notably in the case of overland transport, in which storage of energy in chemical form has always been the norm. In the earliest days of overland transport, storage was biochemical, in the form of horse feed. Subsequently, the principal form of storage was in coal hoppers on trains. These were subsequently replaced by diesel/gasoline storage tanks and, more recently (in some road vehicles anyway), by batteries. We will consider the issues surrounding these alternatives in Chapter 8.

As we move into the other modes of energy use, storage becomes rather more challenging. **Thermal energy storage** is feasible, most obviously in the form of hot water held in large tanks. Underground storage of hot water in permeable rocks is a variant on this concept that is currently receiving increased attention. Other possibilities are the use of '**phase-changing materials**', which change form ('phase') from solid to liquid, or liquid to gas, at temperatures of relevance to thermal system management. Many such phase transitions involve the consumption of significant amounts of thermal energy in the form of **latent heat**, which is the input of heat needed to melt a solid or transform it into a vapour (or to evaporate a liquid) without a change in temperature. For instance, when you heat a pan of water, the rise in temperature will be proportional to the amount of heat supplied until the water reaches 100 °C. Thereafter, continued heating will not raise the temperature further, but will result in evaporation. It turns out that the amount of heat required to do this – the latent heat – is far more than you would expect from the amount needed earlier to warm the water by one degree. Because of this effect, lots of energy can be stored and later recovered by shifting some material within a closed container from one phase to another. But whether thermal energy is stored as hot water or in more exotic phase-changing materials, it is still subject to the Second Law of Thermodynamics: energy wastage occurs as heat leaks out of the store, and the interior cools.

It is when we turn to **storage of electricity** that achieving demand–supply balance becomes especially tricky. It is possible to store electrical charge in chemical form – either in batteries or by using it to produce some useful compound such as hydrogen (which can be obtained from ordinary water by running an electrical current through it, a process called **electrolysis**) – though this incurs energy losses and involves the use of relatively exotic materials that make it a very expensive option at large scale. The other main strategies for electrical charge storage are physical. Some, such as capacitors or flywheels, offer very short-time storage that can be used to smooth the output of generating plant. At larger spatial and temporal scales, the foremost large-scale option is to use spare electrical capacity during periods of low demand to pump water uphill to a reservoir, whence it can later be released to generate power using hydroelectric turbines. This approach, known as **hydroelectric pumped storage**, was pioneered at the Cruachan site in the Highlands of Scotland and has now been adopted at many sites worldwide; it is to this day the only technology capable of delivering gigawatts of power to regional electricity grids at very short notice.

Cruachan: the world's first pumped-storage hydropower station

Originally commissioned in October 1965, Cruachan remains very much in active use: indeed, consideration is currently being given to doubling the capacity of the system. When you first look at the aerial photograph of the site (Figure 4.2), you could be forgiven for asking 'Where's the power station?' The small cluster of buildings at the foot of the hill is used for administrative purposes and is clearly too small to house much in the way of electrical plant. The Cruachan turbine hall is actually entirely underground, one kilometre into the hill from these buildings, accessed by a tunnel. The hall houses four turbines, each rated at 110 MW, which are connected to the dammed upper reservoir (capacity

▲ Figure 4.2 The Cruachan Hydroelectric Pumped Storage Power Station, near Taynuilt, Scottish Highlands.

10 million cubic metres) via a 360-metre vertical shaft (the 'penstock'). A network of high-level aqueduct tunnels bring extra water from beyond the conspicuous peaks of the *Cruachan Beann* mountain range; this helps to defray the need for running the turbines in reverse to pump water up from the lower reservoir – Loch Awe, the large natural lake in the foreground – during periods of low electricity demand (typically in the middle of the night).

Depending on antecedent conditions, the power station can increase its output to the UK National Grid from zero to a full 440 MW in as little as 30 seconds (and seldom any longer than two minutes), and then sustain this output for up to 22 hours, depending on how full the upper reservoir was initially. The operating rules require that the Cruachan system usually retains at least enough water for 12 hours of generation. The peak throughput of water through the turbines is about 200 m^3/s, and the turbines can pump as much as 167 m^3/s to the upper reservoir at other times. (The second figure is lower because of the energy lost in lifting water 360 metres against gravity.) The Cruachan system has been known to produce up to 900 gigawatt-hours (GWh) (= 3.24 petajoules) of electrical energy per year – almost all in response to urgent requirements during periods of peak demand.

Other physical storage strategies for electrical charge are currently being trialled, such as those using air (either compressed into underground cavities, or chilled to around -200 °C so that it liquefies), which can be quickly expanded through a turbine when demand increases. Though showing great promise, none of these technologies is yet capable of coming close to the capacities achievable using hydroelectric pumped storage. As Figure 4.2 illustrates, hydroelectric pumped storage is most effective in areas of high topographic relief – though these also tend to be areas prized for their scenery, which limits the scope for their proliferation in mountainous regions.

With practical limitations on our ability to store electrical energy at large scale and over extended time periods, how else can we balance demand and supply?

▶ Peaks and troughs: baseload and dispatchable power

Human behaviour leads to inevitable changes in demand for electricity, over daily and seasonal cycles. Figure 4.3 illustrates typical patterns observed on the UK National Grid over a 24-hour period.

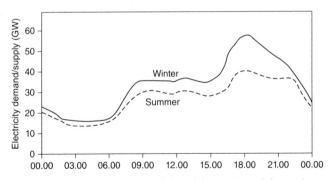

▲ Figure 4.3 Seasonal patterns of electricity supply and demand over one day, UK National Grid

(We could draw a similar diagram for demand and supply of domestic gas, which is predominantly used for space heating, hot water and cooking.) A few moments' reflection on the shape of the curves in this diagram reveals patterns that closely reflect the lifestyles of the majority of humans, with power use increasing after 6 a.m. as everyone breakfasts and scrubs up ready for work or school, then a fairly steady demand through the main working hours (with a shallow peak at lunchtime), before demand rises steeply as everyone cooks dinner and settles down to watch TV or use other recreational appliances.

Given that the troughs and peaks on the above diagram are repeated daily, there is a clear need to match varying demand over time. As we have already seen, electrical power storage has only a limited contribution to make, even where pumped-storage systems are connected to the power grid. For instance, the UK National Grid is currently connected to about 100 sizeable power stations with a total combined peak output ('**installed capacity**') of about 70 gigawatts (GW); hydroelectric pumped storage is supplied by only four stations (with a fifth now planned), together providing a capacity of about 2.86 GW, or about 4 per cent of total capacity. While this storage capacity is useful for short bursts of very high demand, it is clearly nowhere near sufficient to cope with all of the variation in demand experienced by the grid over its daily cycles.

So, if they do not have access to enough storage to do the job, how do the managers of large regional electricity grids cope with fluctuating demand? The answer is by **load tracking**: this means permanently keeping a very close eye on trends in demand – and, so far as this is possible, anticipating them – so that the total amount of power being drawn from the available power stations can be adjusted minute by minute to match demand almost exactly. There are no other real options: to avoid grid instability, which would lead to localized overload and damage, the electronic frequency at which the system is operated must be kept within very tight margins. While there is a little more tolerance for voltage variation, load tracking by bringing power stations on or off line is the only real scope for balancing demand with supply.

What's in a watt?

So far, we have mainly quantified energy in units of joules, which is useful for summing up the total amount of energy available from a certain mass or volume of fuel, or how much energy is expended in a given endeavour or time period. But if we wish to discuss *how quickly* energy is being expended, then we need to specify the number of joules per second. A joule per second is known as a watt, named after James Watt (1736–1819), who increased the efficiency of steam engines to the point that their economics became unassailable. The watt is the unit of power, which is the rate of energy use. Though hardly a difficult concept, it is amazing how many pundits hold forth on the topic of energy while displaying a thorough ignorance of the distinction between energy and power. Just as with joules, the magnitude of power operations is such that it quickly becomes necessary to use the prefixes *kilo-*, *mega-*, *giga-*, *tera-*, *peta-* and so on (see Chapter 2). As if there were not already enough scope for confusion, engineers have got into the clumsy habit of summarizing total energy expended in an hour by multiplying the instantaneous rate of energy use (in watts, or more usually its larger derivatives, especially kilowatts, megawatts and gigawatts) by the number of hours over which it is sustained to produce values of energy used in kilowatt-hours (kWh), megawatt-hours (MWh) and gigawatt-hours (GWh). As a joule per second (i.e. a watt) multiplied by a number of seconds simply gives us joules again, it would be neater to just use joules (or kJ, MJ or GJ as appropriate) for such purposes – but the genie won't go back in the bottle any more. But in all cases the conversion uses the same factor of 3,600 – that is, there are 3,600 kJ in 1 kWh, 3,600 MJ in 1 MWh, and so on.

As we shall explore in greater detail in the next chapter, not all energy conversion technologies lend themselves to being turned up and down, let alone switched on and off, at short notice. Recognizing this, certain power stations are kept running at pretty much constant output, to meet the 'irreducible minimum' of power demand, while other more flexible stations adjust their output to match the changing demand profile. The constant output stations are said to be providing '**baseload**', while those that vary their output to cope with the peaks in demand are said to be providing '**dispatchable**' power. Figure 4.4 shows how these two components of supply are deployed through the 24-hour demand cycle:

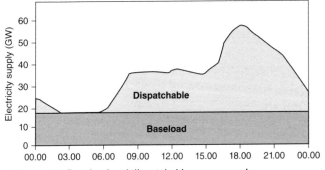

▲ Figure 4.4 Baseload and dispatchable power supply.

Since this load-tracking approach first evolved in the first half of the twentieth century, it has served the electrical power supply needs of the industrialized, urbanized countries very well. Innovations in electrical energy supply technologies were devised with this approach in mind. At present, nuclear and coal-fired power stations are the principal suppliers of baseload in most OECD countries, with gas- and coal-fired stations supplying dispatchable power.

At the turn of the twenty-first century, as the imperative of addressing climate change began to be translated into political commitments, other priorities came to the fore in energy technology development: minimization of carbon emissions has become the overriding consideration for most energy-sector innovators. It is not difficult to spot where the difficulty arises in this scenario: given that power is produced by wind turbines only when wind conditions are suitable, and that solar panels produce power only during hours of daylight, these most widely deployed renewable technologies offer *neither* baseload *nor* dispatchable capabilities. Rather, they are available for our use solely at the largesse of Nature. While there are some renewables that can offer baseload (most notably mid- and high-enthalpy geothermal, biomass combustion and hydropower), and even some degree of dispatchability (biomass and high-enthalpy geothermal), it is unlikely that these are going to be available at sufficient scale to significantly displace fossil-fuelled generation for at least several decades. Herein lies one of the greatest challenges of the modern era: how to minimize carbon emissions without ending up with power cuts. To get a good grip on our options, we need to appreciate the capabilities and limitations of the energy conversion technologies available to us. The following two chapters consider these.

5

Upstream activities: preparing for energy conversion

'Togaidth mi mo theine an diugh,

An lathair ainghlean naomha neimh [...]

Gun ghnu, gun tnu, gun fharmad.

Gun ghiomh, gun gheimh roimh neach
fo'n ghrein, [...]'

(I will kindle my fire this morning,

In presence of the holy angels of heaven [...]

Without malice, without jealousy,
without envy.

Without fear, without terror of anyone
under the sun, [...])

Anonymous, Scottish Gaelic

▶ Conversion, not generation

As we have seen, energy can neither be created nor destroyed; it can only be converted from one form to another. This is precisely what the First Law of Thermodynamics makes clear. It is therefore simply wrong to talk about 'generation' of energy/power. This is not mere pedantry: it encourages consideration of how to make the most of limited energy resources. For instance, it has been the traditional practice of the electricity-generation sector simply to dispose of any thermal energy too cool to spin a turbine. This is why large fossil-fuelled and nuclear power stations always have large cooling towers emitting steam – the 'waste' heat is simply discharged to the atmosphere. And yet there are myriad uses to which this 'waste' heat could profitably be put such as **combined heat and power (CHP)** developments. The advantages of taking a CHP approach are clear: a traditional coal-fired power station will rarely achieve an efficiency of energy conversion in excess of 40 per cent (values in the mid-30s are common), whereas a CHP system can make use of 80 per cent of the energy converted. This, in turn, means that, even where a CHP system uses a very high-carbon fuel such as coal, its carbon emissions per joule of useful energy supplied will be half those of a conventional coal-fired power station.

It is clear from this example that decisions taken quite early in the development and exploitation of a potential energy resource can have important implications for later flexibility in energy use. Let's examine some of the preparations that need to be made prior to accomplishing the final conversion of energy into a useful form. Depending on the natural energy source, minimal intervention might be needed to deliver the energy in a convenient form. In the days of steam trains, for instance, coal was simply dug up and stored on board to supply the fires that turned water into steam. A more contemporary and far lower-carbon example of energy use following minimal conversion is provided by **geothermal** wells, many of which produce hot water that can be directly used in district heating networks. This is the case, for instance, in the Icelandic capital, Reykjavik, where about 490 megawatts of thermal energy (MW_{th}) are obtained directly in the form of hot water pumped from aquifers beneath the urban area. In such a case, preparation for energy conversion simply amounts to pumping naturally hot water from wells and distributing it in pipe networks.

▶ Preparing conventional fossil fuels for use

If heat delivery can be quite direct, increasing complexity characterizes the work needed to prepare for production of transport fuels and electricity. Moving up a level of sophistication from an antique steam engine, **coal preparation** for modern power stations typically involves washing the mined coal to remove any extraneous rock and soil fragments, crushing the washed coal to obtain particles of a size suitable for combustion, dewatering the crushed coal, then transporting it, storing it and finally delivering it to the furnace.

Both oil and natural gas are, of course, ultimately obtained from natural geological reservoirs (basically layers of porous rock) using borehole technology. In many cases, these rocks are sufficiently permeable that they will readily yield their hydrocarbons to boreholes. The hydrocarbons obtained from such wells are termed conventional. Where the rocks are less obliging, various technologies are used to stimulate their permeability. Although regarded as a recent innovation, hydraulic fracturing (or 'fracking') has been used to enhance the permeability of reservoir rocks of various types for more than half a century. So-called unconventional gas is simply natural gas obtained from a type of rock that would not in the past have been regarded as an economically viable reservoir – such as shale or low-permeability coal seams.

Transforming crude oil into the principal liquid transport fuels (i.e. gasoline/petroleum and diesel) is one of the principal activities of an oil refinery. This is achieved by a process known as 'fractional distillation' (Figure 5.1), in which the incoming crude oil is heated, and passed through columns in which the various components of the oil separate out on the basis of their different boiling points. In this way, a whole range of fuels can be produced from a single precursor material – though, in reality, some crude oils are better suited to gasoline the production, while others are better suited to the production of diesel or fuel oil.

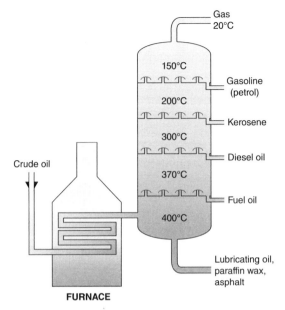

▲ Figure 5.1 A simplified summary of the separation of crude oil into different hydrocarbon fuels by means of fractional distillation in an oil refinery.

While **natural gas** can be recovered from hydrocarbons during fractional distillation, about three times as much is produced from gas-only boreholes accessing reservoirs that do not also contain appreciable oil. This natural segregation of hydrocarbons largely reflects the different temperatures at which sedimentary organic matter is transformed into oil (60–120 °C) and natural gas (100 – > 200 °C) – though the overlap between the two temperature ranges partly explains why both occur together in many reservoirs. Processing of gas from gas-only wells involves separation of the predominant gas, methane (CH_4), from:

▶ other hydrocarbons that can be sold separately, such as ethane (C_2H_6), propane (C_3H_8) and butane (C_4H_{10})

▶ unwanted gases, such as water vapour and the 'sour gases' carbon dioxide and hydrogen sulphide.

Once scrubbed of such impurities, and therefore also dried, natural gas can most easily be transported to end-users by pipeline. For longer-distance transport, chillers can be used to condense methane into the much denser form of **liquefied natural gas** (**LNG**), though this imposes a substantial energy penalty, resulting in far higher costs than pipeline gas. (This is why, to date at least, natural gas has traded in regional markets, rather than in the unified global markets which have long governed sales of oil and coal).

▶ Bioenergy preliminaries

Had this book been written only a decade ago, there would have been little cause to mention preparation of biomass. After all, most humans through history have prepared wood for the fire and so are aware that it needs to be cut small enough to fit in the grate and to be dry enough to catch flame without undue delay. Now that large-scale combustion of woody **biomass** is favoured by many governments as an alternative to coal-fired generation, there is a need to deploy the same approaches at much larger scale. Fresh timber commonly contains about 50 per cent water by mass, and it is common to reduce this to 15 per cent or less to increase flammability.

Increasingly, woody biomass is mixed with coal for combustion in conventional boiler plants. For this application, it is desirable to try to close the substantial gap in energy density between wood and coal. Drying helps in this process, but it is necessary to drive off other volatile compounds to maximize the energy density of the wood. This is typically achieved by a process known as **torrefaction** (literally 'tower building'), in which stacked wood is heated for about half an hour at temperatures in excess of 200 °C. As the bulk density of wood is also considerably lower than that of coal, it is desirable to increase this to expedite transport and reduce space requirements for stockpiling. This can be achieved by grinding the dried (and often torrefied) wood chips to powder and then reconstituting this powder into dense pellets of approximately uniform diameter.

Similar processes are commonly applied to other forms of solid biomass, though some of these require even more pre-processing: for instance, to prepare **municipal waste** for combustion, it is also necessary to remove non-combustible components (glass, metal) or those with potential to release toxins (e.g. certain plastics). This separation is achieved using an array of gravitational and magnetic techniques adapted from the mineral-processing industry. The remaining material typically comprises about 85 per cent biomass, and emerging technologies such as rotational **autoclaving** can convert most of this into sterile cellulose fibre, which can either be pelletized for combustion or used as a feedstock, alongside sewage sludge and farm waste, for production of **biogas** by means of anaerobic digestion.

Anaerobic digestion harnesses in industrial form the natural microbial fermentation processes that occur in many settings, such as in marshes, in the digestive systems of mammals (notably ruminant cattle, but also in humans) and also in landfills. A particular group of microbes, which experience oxygen as a systemic poison, absolutely thrive in the absence of air (hence 'anaerobic'). These derive their energy by breaking down natural organic molecules such as acetate – previously derived from cellulose by composting processes involving other, *aerobic* bacteria. Once the organic matter is ready for anaerobic digestion, it is typically placed in a large reactor tank, from which air is usually excluded simply by ensuring submergence of the biomass in water (albeit dry digestion is also achievable in specially designed plants). Anaerobic digestion occurs most efficiently at elevated temperatures, so the high specific heat capacity of the water is a disadvantage in this regard, as the water requires heating. The overall process of anaerobic digestion typically results in a halving of the solids content over a cycle of 48 hours or so, with the residual solids being useful as soil amendments, restoring valuable carbon to farmland; they can also be dried, pelletized and used for biomass combustion. Anaerobic digesters produce a gas that typically comprises two-thirds methane to one-third carbon dioxide. (Biogas evolved from landfills also contains about 15 per cent nitrogen, and concomitantly less methane – around 45 per cent.) Because of the dilution of methane by CO_2 or nitrogen, biogas usually has a rather lower energy density than natural gas obtained from geological reservoirs: where natural gas will typically have about 40 MJ/kg, biogas from anaerobic digesters is normally closer to 23 MJ/kg, while landfill biogas usually has about 12 MJ/kg.

Biogas from anaerobic digestion of sewage: how much energy can it produce?

Take one of the UK's finest examples of anaerobic digestion: the Bran Sands plant operated by Northumbrian Water Ltd in north-east England. The anaerobic digesters receive all of the sewage sludge from an enormous wastewater treatment plant that serves a population equivalent to 1.2 million people. You might imagine that the gas produced by the anaerobic digester would make a significant dent in the energy needs of that population. In fact, as correctly anticipated by the design engineers, the gas yielded by the digesters is used to produce some 4.3 MW of electricity – which is sufficient to supply up to 70 per cent of the electricity demand of the sewage works, but can make no net contribution to the energy needs of the population the works serve.

There are other processes for obtaining methane from biomass, most notably pyrolysis and/or gasification, in which biomass is heated in an oxygen-free or oxygen-sparse environment respectively. The resultant gas is usually called **synthesis gas** or **syngas** (a rather loose application of this term), and it principally comprises hydrogen, carbon monoxide and methane, with variable amounts of carbon dioxide. The residual solid produced during biomass pyrolysis is a form of charcoal known as **biochar**, which, like anaerobic digester residues, has great value as a soil amendment: indeed, depending on the source of biomass used, this can be a '**carbon-negative**' process, which actively removes CO_2 from the atmosphere.

It is often blithely assumed that bioenergy produced from farm wastes or sewage is 'green' and low-carbon. This somewhat flies in the face of the fact that the vast bulk of groceries and animal feed consumed worldwide today is produced by intensive agriculture that depends on the application of huge amounts of nitrogenous fertilizers – which are overwhelmingly produced using fossil fuels. Despite all the best efforts of the organic farming movement to date, we are nowhere near having a comprehensive alternative to fossil-fuel-based fertilizers, and, until we do, the low-carbon credentials of many supposedly 'green' biomass energy projects are open to question.

Nuclear resources: enriching uranium

Mining and **processing of uranium** ores are the essential prerequisites for production of nuclear energy. We have already seen that attainment of nuclear fission requires substantial enrichment of the ^{235}U content of uranium, though, even before this can be done, natural uranium ores (in which the ^{238}U isotope predominates) have to be extracted from enclosing 'gangue' minerals, ground down fine enough to liberate individual grains of the uranium minerals, then separated from other grains using an array of density and/or gravitational sorting techniques. In the process, large volumes of fine-grained mineral waste ('tailings') are produced, which require careful handling to avoid risking dangerous mudflows and/or pollution by airborne dust or dissolved metals. The richest uranium ores ever mined in any quantities have ranged up to about 20 weight per cent of uranium (wt % U). At present, ores containing as little 0.02 wt % U are being mined, though most mines in free-market economies operate above 0.1 wt % U. Although the classic principles of economic geology suggest that declining ore grades will push up prices and thus drive innovation in the recovery of ore from ever-lower-grade orebodies, in the case of uranium an irreducible energy threshold exists, below which the amount of energy needed to recover the uranium will exceed the amount of nuclear energy that it could ever produce. Current consensus suggests that this threshold lies between 0.015 and 0.010 wt % U. Bearing this in mind, and allowing for future orebody discoveries at rates comparable to those of the past half-century, there are currently between 50 and 70 years of uranium resources still to mine, at current rates of consumption. This is one of many reasons why a switch to thorium-based nuclear reactors is being explored.

Thorium has many other factors in its favour, including its greater natural abundance; an energy cycle in which the element is recycled many times so that waste is minimized; wastes that remain radioactive for only a few decades rather than millennia; its intrinsic safety when used in molten salt reactors; and no risk of diversion into weapons by anyone – governments or terrorists.

Beyond the fission chains supported by uranium and thorium (which in itself is *fertile* rather than fissile),

nuclear fusion remains the holy grail of nuclear energy. Although fusion (the merging of nuclei rather than splitting them) is the source of all solar radiation, it has yet to be emulated in a sustainable industrial process capable of producing more energy than it consumes. In the Sun, fusion takes place at enormous temperatures, and the amount of energy needed to emulate this on Earth requires prohibitively expensive preheating. If fusion can eventually be achieved at lower temperatures (so-called **cold fusion**), the rewards for humankind would be enormous: abundant energy from effectively limitless raw materials with no appreciable waste legacy. However, throughout my career (which has already lasted more than 25 years), I have repeatedly been assured that cold fusion will be an industrial reality in 25 years' time ...

▶ Renewable resources: plug and play?

Preparations for conversion of energy from wind, wave, solar and tidal sources contrast markedly from those for combustibles and nuclear resources. For the most part, the preparations for harnessing these renewable sources amount to a combination of direct measurements and calibration of mathematical models to characterize:

▶ meteorological parameters, such as the intensity of solar radiation, wind speeds and directions, and wave frequencies and amplitudes

▶ tidal ranges and current velocities.

Thus, for instance, the first substantive activity in the development of a new wind farm is erection and sustained monitoring (preferably over a number of annual cycles) of a 'met mast' – that is, a mast fitted with meteorological instruments, such as a wind vane and anemometer. In countries with well-established wind- and solar-power sectors, the results of many such investigations have been collated by government agencies into online atlases that can provide predictions of design weather parameters for any specified location.

Geothermal energy stands in a category of its own when it comes to preparing for energy conversion. Exploitation of hot groundwaters demands the development of geological reservoir models, which in most respects resemble those developed for oil and gas reservoirs – with the exception that reinjection of spent water sustains pressures in geothermal reservoirs over extended time periods, with recirculating water heating up as it passes through the strata. Reservoir characterization proceeds by means of exploratory boreholes, test pumping of pilot boreholes, interpretation of the temperatures and chemistries of produced waters, and mathematical modelling of natural and induced movement of water through the strata.

6

Energy conversion technologies

'1. You can't win

2. You can't break even

3. You can't get out of the game.'

Allen Ginsberg

Having prepared our energy resources for conversion, we finally obtain usable energy through the application of a range of technologies. These vary significantly between the three modes of energy use.

▶ Heat: from hearth to heat pump

For millennia, humankind relied almost entirely on wood as a heating fuel. To this day, wood is the sole heating fuel in many developing countries, with as much as 80 per cent of all wood harvested in some African countries being used as fuel, and a similar proportion of all energy use in those countries accounted for by **firewood.** A debate rages as to how sustainable this practice is, but many African countries uncritically report domestic firewood use in their renewable energy statistics. In doing so, they take note of the move back to firewood as a supposedly 'sustainable' fuel in prosperous countries. We are therefore living through a weird era in which the use of wood as a primary heating fuel is increasingly restricted to the very poorest and very richest people in the world, with few in between realistically able to avail themselves of it.

It was not always thus, of course. In what are now the industrial (or post-industrial) countries of northern Europe and North America, dwindling wood supplies during the eighteenth and nineteenth centuries led to wholesale displacement of wood by **coal** as the domestic heating fuel of choice. However, coal smoke is more problematic than wood smoke. Hence in the twentieth century, as gas became readily available from centralized town gasworks (in which coal was gasified), and later from natural gas reservoirs, **gas-fired central heating** and cookers became the predominant option, being both cheaper and cleaner than coal. Only in rural areas not served by gas grids are coal or **fuel oil** still used for domestic heating. The latter are not only more expensive: they are also far heavier in carbon emissions than natural gas.

Irrespective of the fuel source, almost all delivery of heat takes place through the medium of water or air. This involves transfer of heat from the furnace source into a carrier fluid (or a sequence of such fluids) by means of a **heat exchanger.** There are many different designs of heat exchangers, though all are essentially variants on the same basic concept of juxtaposing the hot gas supplying the heat with the fluid to be heated across an impermeable but thermally conductive interface (usually made of metal). The fluid to be heated is usually termed a secondary **working fluid.** Once the working fluid has been heated, electrical pumps are generally used to force it through pipework to deliver thermal energy to end-users.

A further nuance of the concept of working fluids arises in relation to **heat pumps**. Whether they yet realize it or not, all readers of this book are familiar with heat pumps: they are the technology used to chill the interiors of refrigerators and freezers. While you have most likely paid attention to the contents of their cool interiors, you have probably noticed that the backs of these appliances tend to be very warm. This is because the heat pump extracts thermal energy from the air inside the refrigerator/freezer and expels it into the air outside it, warming it in the process. You are probably also aware of the sound of machinery spinning that sometimes comes from the back of a refrigerator/freezer, particularly after you have had the door open for a while so that the interior has become warmer than desired. This mechanical noise is from the operation of a compressor, a device that squeezes a working fluid through tubes between the interior of the refrigerator/freezer and the rear exterior of the cabinet.

Remembering schoolbook physics, the 'Ideal Gas Law' tells us that increasing the pressure of any gas will increase its temperature: just think of how hot the tip of a bike pump adapter can get when you pump up a tyre. Heat pumps use this effect to transfer heat from a cool place to a warm place. For heating applications, the 'cool place' is typically water (e.g. a lake, stream or well), the ground (either loose soil or bedrock accessed via boreholes), or simply the air, and the 'warm place' is the interior of a building. Depending on the source used, heat pumps tend to be termed '**ground-source heat pumps**' or '**air-source heat pumps**'. As the air outside is usually at its coldest during the season with highest heat demand, while the subsurface typically maintains a fairly constant temperature year-round, ground-source heat pumps are generally preferable from the point of view of energy efficiency and minimization of associated carbon emissions. It is important to realize that heat pumps are quite different from conventional gas or electric heaters, in which air is heated by passing it over gas flames or a metal coil being heated by electrical resistance: where a heat pump is working as designed, more than three-quarters of the thermal energy that it delivers will be derived from the natural environment and will ultimately be mainly solar in origin. If the electricity used to operate the heat pump comes from a low-carbon source, then it is a thoroughly low-carbon and renewable source of heating.

In some cases, ground-source heat pumps intercept at least some genuine **geothermal energy**, and, as we have seen, direct use of hot groundwaters for district heating is an attractive source of renewable heat in many regions. Though developed mainly in volcanic regions to date, there is abundant evidence that medium-enthalpy geothermal reservoirs with temperatures suitable for heating purposes are widespread, and could renewably supply a large proportion of heat demand at mid to high latitudes.

At low and mid latitudes, **solar hot water** is already a success story: many dwellings in tropical latitudes obtain all their domestic hot water from this source, and in such settings thermal convection and gravity alone can circulate the hot water from roof-mounted storage tanks to taps and shower heads without any need for auxiliary pumping. Sadly, solar thermal yields are not sufficient to meet winter space heating needs in places such as northern Eurasia, North America, Patagonia and New Zealand. Nevertheless, even at these high latitudes, solar thermal systems can typically provide all hot water needs from late spring through to early autumn, albeit with the aid of electric pumps.

▶ Transport: from horsepower to hydrogen

The most fundamental source of transport energy is human muscle power, ultimately sustained by food, of course. If we wish to develop truly sustainable and healthy urban lifestyles, we need to be more creative about recreating opportunities for using human muscle power in transport, especially by walking and cycling. Other modes of transport using muscle power, such as rowing, are also of great recreational and sport value, though seldom used seriously for transport any more. There are, though, serious limitations on how much one human being – or even a whole team of them – can transport, both in terms of cargo load and speed. Although world-leading cyclists in the Tour de France can sustain peak power outputs of as much as 400 watts for an hour or so, few lesser mortals would manage to sustain much more than 75 watts throughout a full working day.

It has long been recognized that large mammals are far more powerful than humans, with horses being capable of power outputs typically ten times that of humans. With the exception of wind propulsion of boats, horse-drawn transport was the only serious alternative to human muscle power until the advent of steam engines in the 1820s. Although horse-drawn transport is largely obsolete nowadays, many people still ride horses for leisure, and nostalgia for this form of transport endures. Perhaps this explains why many people still choose to express power in units of '**horsepower**' rather than watts. This tendency is strongest in the United States, which built a large part of its national identity around the image of horse-riding cowboys. In case you are wondering, there are 745.7 watts in 1 horsepower – in other words, just enough for the efforts of a single horse to power your hair dryer in the morning (though don't try that at home if you care about your bathroom decor).

George Stephenson (1781–1848) first successfully mastered **steam locomotive** technology around 1820. Steam engines had been around long before then but had principally been used as fixed engines operating pumps and cable haulage in the UK coal-mining industry. Richard Trevithick (1771–1833) pioneered the concept of the steam locomotive in 1802, but a number of technical challenges had to be overcome before the technology became cost-competitive with horse-drawn railways. There are many valuable lessons to be learned from the history of steam locomotion, but one of the most relevant to the twenty-first-century energy debate is this: in millennia of using wood to heat water, no one had ever managed to produce a worthwhile *wood-fired* locomotive. This is because, as we noted in Chapter 3, both the power density and energy density of wood are far lower than that of coal, which means that combustion of a lot more wood than coal is required to perform the same amount of work; furthermore, elevated temperatures are far easier to achieve with coal than with wood. This historic insight has obvious implications for the modern trend of converting formerly coal-fired power stations to wood-firing.

As you probably know, in a steam locomotive coal is burned in a grate beneath a boiler, in which water is heated to produce high-pressure steam. The steam is then used to move pistons up and down in a cylindrical chamber within the engine. As the combustion of fuel takes place outside the engine cylinder, a steam locomotive is an example of an *external* combustion engine. One disadvantage of this is that much of the energy is wasted, simply exiting the smokestack. It wasn't long before it occurred to people that greater use could be made of the available energy if the combustion occurred *within* the engine cylinder. This is a non-starter for solid fuels such as coal, but as town gas (produced by gasification of coal) became widely available in the second half of the nineteenth century, the way was clear for development of the internal combustion engine. In such an engine, the fuel is mixed with air, forming a highly flammable vapour that is readily ignited by means of an electrical spark. The key to success was devising efficient sequences of events that ensured rapid piston oscillation, which is converted into rotary power by means of an eccentric crankshaft. Originally developed for static applications, it was in the 1880s that internal combustion engines were adapted to use easily transportable liquid petroleum as a fuel, which unlocked their potential as automobile engines. These developments were the fruit of German engineering excellence, and the names of some of the pioneers remain famous in the automotive world today: Gottlieb Daimler (1834–1900) and Karl Benz (1844–1929).

Yet another German engineer, Rudolf Diesel, not only devised a far more efficient internal combustion engine that was to transform the economics of long-distance freight haulage, but also gave his name to the relatively heavy oil-derived fuel that proved optimal for operating his engine. In a **diesel engine**, the Ideal Gas Law is again applied, with air being compressed so that it becomes very hot, at which point fuel is injected and combustion occurs spontaneously, without the need for an electrical spark. To date, the diesel engine continues to reign supreme in heavy haulage by land or sea. It seems highly unlikely that this situation will change radically any time soon. Only two options are currently available: substitution of biodiesel for mineral diesel, and conversion to gas turbine engines supplied with liquefied natural gas (LNG). At present, substitution with biodiesel can only be partial, due to feedstock constraints. Expanding the feedstock base, for instance to include lipids obtained from algae cultivated in seawater, might improve this situation, though it is highly unlikely to provide a total solution. Adoption of LNG would require substantial refitting of engines and the ubiquitous development of new refuelling infrastructure; these factors militate against uptake – an example of **technology lock-in** – even though LNG is already a cheaper fuel than diesel in many regions, and also has far lower carbon emissions.

Among non-fossil and non-combustion options are nuclear, solar, wind and hydrogen. Although **nuclear reactors** have long been used by military submarines, their use in merchant shipping would be fraught with safety and security challenges. Even using the most optimistic projections for improvements in **solar energy** conversion efficiency, the limited surface area available on board ships means that solar photovoltaic technology will never be able to make more than modest auxiliary contributions to shipping energy demands. On the other hand, **wind propulsion** was the prime mover of ships until the advent of steam turbines, so there is no doubt that it can play a significant role. Indeed, many people still enjoy the pleasures of traditional sailing for recreational purposes. However, modern safety requirements would preclude having wind as the sole source of propulsion for large commercial vessels, as it is no longer acceptable for crews to be imperilled should winds fail in mid-ocean. Nevertheless, both fixed and kite-like fabric sails, and deck-mounted rotors (which operate in accordance with a fluid mechanics principle known as the 'Magnus Effect'), can provide substantial auxiliary power with relatively little use of valuable deck space.

Use of **hydrogen** produced by electrolysis of water using renewable energy is perhaps the most promising longer-term option for decarbonizing heavy goods transport. If $E = m c^2$ is the world's most familiar equation, then surely H_2O is its best-known chemical formula. Electrolysis involves the use of electricity to separate ('lyse') water molecules into the constituent molecules of hydrogen (H_2) and oxygen (O_2). This can be written:

$$2 H_2O \Longleftrightarrow 2 H_2 + O_2$$

The two-way arrow shows that the reaction is reversible: pass an electrical current through water and the reaction moves to the right, producing hydrogen gas that can be stored for later use. React hydrogen gas with oxygen (abundant in the air, of course) and the reaction moves to the left, producing electricity. Hydrogen can also be obtained from fossil fuels by gasification. This offers the intriguing possibility of using fossil-derived hydrogen to support the development of infrastructure that in the future can be wholly renewably supplied.

While hydrogen can be burned in conventional combustion engines – indeed hydrogen-fired engines are widely used in space rockets – its direct conversion to electricity is best achieved within a **fuel cell**, which is an electrochemical reactor that produces DC electrical power. Fuel cells closely resemble batteries, but while a battery is a sealed unit containing all the chemical reactants it needs, a fuel cell is fed with reactants ('fuel') on one side, and it releases an exhaust at the other. Although fuel cells can be devised to operate with a wide range of fuels, including alcohol, in the case of hydrogen fuel cells they are fed pure hydrogen and air, and emit water vapour. They are thus extremely clean at the point of use, and, if the hydrogen is sourced renewably, then hydrogen fuel cells provide renewable transport energy. Fuel cells have an advantage over batteries in that refuelling requirements are very similar to those for conventional gasoline. Hydrogen fuel cells are already in use for terrestrial transport, with successful applications in motorbikes, cars and buses. As for LNG, however, the need to develop ubiquitous infrastructure is hindering uptake. More formidable barriers for larger-scale applications relate to technical challenges in providing sufficient energy storage, as, although hydrogen has a high specific energy (~120 MJ/kg), it has a relatively low energy density (10 MJ/litre) (the corresponding figures for gasoline are 45 MJ/kg and 35 MJ/L). The low energy density means that rather large storage volumes are required to ensure sufficient fuel is available for a long journey; on the other hand, the fuel is so light that it imposes little extra load on the vehicle.

Vehicles powered by fuel cells have electric motors. Alternative ways of getting power to electric motors include live rails or overhead cables – as used by trains – or on-board batteries. Electrification of railways is the apex of a transition from horsepower through steam power to diesel power. The initial rationale for electrification of railways was to increase acceleration and thus shorten journey times by 10 per cent or so. An ancillary benefit was to improve air quality by eliminating the particulate emissions associated with diesel engines. Where the electricity used to supply the railway is renewably generated, electrification also serves to reduce carbon emissions.

Conversely, the introduction of electric road vehicles is now being promoted precisely to meet the imperatives of emissions reduction, with speed and range being relegated to secondary considerations. Whether battery-powered vehicles actually make much contribution to emissions reduction depends entirely on the source of electricity used to charge them: where the power is from renewables (rare) or from gas-fired power stations, then significant savings can be made. Coal-fired power stations result in net vehicle emissions no better than those of conventional gasoline or diesel engines.

Unlike trains, cars cannot obtain power from rails or cables, and thus need to carry their electricity source on board. Apart from hydrogen, the principal option is batteries. The traditional lead-acid batteries still used to operate the starter motors of gasoline-fired engines are unsuitable for providing the main engine power source, not least because they are very heavy. Lithium-ion batteries are thus the present technology of choice, because lithium is far lighter than lead, and the resultant batteries use a series of chemical reactions that are far less prone to developing the wasteful 'memory effect' that rapidly reduced the usable capacity of earlier types of battery. However, the batteries of current commercially available electric cars can support journeys of only up to 125 miles on a single charge, and a full charge still takes eight hours (although the process is not linear, and an 80-per-cent charge from flat can be achieved in just 30 minutes). Thus, the present generation of electric cars are unsuited to lengthy journeys but could be suitable for routine commuting runs. However, given that train travel is 40 to 60 per cent more efficient in energy and carbon emissions per passenger mile than car travel, electric cars would not be the first choice for the conscientious commuter. Whether future commuters travel by battery-powered cars or electrified trains, it seems that ever-closer interactions are set to characterize the relationship between transport energy and electricity.

Electricity: a preponderance of turbo-machinery

The discovery and eventual mastery of electricity was a prolonged affair involving many key players, from Alessandro Volta in the late eighteenth century, through James Joule, Georg Ohm and Michael Faraday in the early nineteenth century, to James Maxwell and many others in the latter half of the nineteenth century. This was precisely the same period in which steam engines were being perfected and widely used for static and transport haulage purposes. It was therefore natural that static steam engines were soon harnessed to drive early electrical power generators. In parallel, turbines were also being developed to generate hydroelectricity, with individual power stations achieving outputs of several megawatts by the turn of the twentieth century. However, the real revolution in power generation came in 1884, when previous hurdles to spinning turbines using high-pressure steam were overcome. To this day, the turbine has not been replaced as the fundamental technology of energy conversion for purposes of electricity production.

Turbine technology is at the heart of power-generation systems using coal, gas, biomass, nuclear, geothermal, concentrated solar power, wind, hydro, tidal and (in many cases) wave resources. Where steam engines and internal combustion engines use reciprocating pistons within cylinders to produce motive force, turbines are rotating devices that can spin a shaft which (with suitable acceleration using gears) can directly supply the rotary action needed to operate standard electrical generators.

While the basic principles of turbine technology can be traced back to ancient Egypt, it was not until 1884 that the technology was mastered for the purposes of energy conversion. This key innovation was the milestone achievement of Charles Parsons (1854–1931), and he cheekily demonstrated its potential by gate-crashing a naval regatta on the occasion of Queen Victoria's Diamond Jubilee, held off Portsmouth, UK, in 1897. Parsons' boat, *Turbinia*, which was the first turbine-powered boat in the world, proved to be easily the fastest boat in the world at the time, running rings around naval pursuit vessels that were dispatched to arrest it. The Royal Navy soon recognized the value of the invention, however, and within two years it had commissioned turbine-powered warships of its own. Within three decades, the experience gained with steam turbines gave rise to the first highly effective designs for **gas turbines**, or **jet engines** as they are now more commonly known. Gas turbines are the logical extension to the world of rotating machinery of the prime principle of internal combustion engines. While jet engines have proved to be the defining energy conversion technology for the modern aviation industry, both steam and gas turbines remain the principal means of energy conversion in electrical power production worldwide.

Although the details of power generation using turbines are complex, the principles are simple enough: a high-pressure stream of fluid is passed through several banks of inclined blades attached to a single shaft (or axle), so that displacement of the blades causes the shaft to rotate at high speed. The shaft transmits rotary force (or 'torque'), via a series of gears, to a neighbouring shaft, which rotates within the cylindrical housing of a generator. The gears ensure that the rotational speed of the generator shaft is ideally tuned for the final stage of the process, in which devices (either brushes or magnetic coils) are rotated within an external applied magnetic field, giving rise to a flow of electrons, which constitutes an electrical current. Although there are numerous variants of fluid type (water, air, steam, combusting gas) and generator configuration (e.g. brushed, brushless, permanent magnet), these basic principles apply to all turbine-based electricity generators.

Detailed design of a turbine for a specific purpose relies on rigorous thermodynamic analysis of the process used to generate the high-pressure fluid and its progress through the turbine. Most power stations fired using fossil fuels or biomass, and all nuclear and geothermal power stations, harness turbines by means of the **Rankine Cycle**, named after the eminent Glasgow University engineer William J.M. Rankine (1820–72). It is impossible to overstate the importance of the Rankine Cycle: at present it is responsible for production of around 85 per cent of all electricity used worldwide. In the Rankine Cycle, water is boiled to produce high-pressure steam, which is then used to spin a turbine, after which it is cooled and returned to the liquid form by passing it through a type of heat exchanger known as a condenser. In this case, the combustion (or other heat source, such as a nuclear reactor) is *external* to the turbine, and the working fluid is water. A separate flow of cold water is passed through the condenser receiving the steam exiting the turbine. If you have a heat source that is insufficient to raise high-pressure steam, it can still be harnessed to produce electricity using a variant of the Rankine Cycle in which some substance with a far lower boiling point than water is used as the working fluid, with a heat exchanger being used to transfer thermal energy to it from the primary source. Quite which substance is used depends on the temperature of the primary heat source, and the specifics of plant design depend in turn on the nature of the selected substance. Usually, an organic compound is used for the secondary working fluid, and the resultant power production facility is said to be operating according to the **Organic Rankine Cycle**. An alternative system using a mixture of ammonia and water as the secondary working fluid is known as the **Kalina Cycle.** These variations of the basic Rankine Cycle have vast potential to harvest electricity from medium-enthalpy geothermal fluids, biomass combustion plants, concentrated solar power plants (in which solar heat is used to boil water), and from waste heat produced in many industrial processes. Because they involve using both a primary heat source and a secondary working fluid, these systems are also known as **binary cycle plants.**

A typical coal-fired Rankine Cycle power plant will convert around 30 per cent of the energy released from the coal into electricity. Gas turbine plants, which operate according to the **Brayton Cycle**, in which combustion is *internal*, are much more efficient, and it is common for the waste heat they produce to be used to raise steam, which is then converted to electricity using a second turbine operating according to the Rankine Cycle. The combination of Brayton and Rankine Cycle turbines in series is referred to as a **Combined Cycle Gas Turbine (CCGT)** plant, and they typically achieve energy conversion efficiencies approaching 60 per cent. Since gas is in any case far less carbon-intensive than coal, CCGT efficiencies thus result in substantial decarbonization of electricity production, despite the fact that natural gas is also a fossil fuel. Even greater results are obtained when a CCGT plant is operated in combined heat-and-power (CHP) mode, with waste heat being captured for other uses: overall energy efficiencies can then exceed 80 per cent.

Nuclear power stations use the vast amounts of thermal energy released from subatomic particles in accordance with $E = mc^2$ by heating water to steam, which is then passed through a turbine. The water does not come into contact with the nuclear reactants; all of the heating takes place using heat exchangers.

The applications of turbines discussed above are all **thermal**, regardless of the fact that some are renewable (geothermal, biomass) and some are based on fossil carbon (coal, gas) or uranium (nuclear). Yet turbines are also key to many other energy conversion processes. In fact, some of the earliest turbines were those developed for **hydropower** purposes. Although waterwheels have been used since ancient times, they tended to use devices with rotational axes perpendicular to the direction of water movement. Only the **Archimedean screw** rotated in an axis parallel to the flow direction, though the historic use of that device was for pumping rather than power production; it is only recently that it has found use in low-head hydropower systems. It was in 1878 that modern hydropower emerged, with a small turbine being used to generate electricity at the home of the great industrialist Lord William Armstrong at Cragside, Northumberland, UK. Much larger installations soon followed in North America. The power output of a hydroelectric station of any size is fundamentally governed by the **flow rate** (Q) of the available water and its **head** (h) – that is, the height through which it drops as it flows through a pipe (known as the **penstock**) to the turbine. Leaving aside energy losses due to friction in the penstock and mechanical inefficiencies in the turbine, the theoretical maximum power output (P) of a hydropower station is easily calculated using this formula:

$$P = 9.81Q\,h$$

where P is in kilowatts, Q in cubic metres per second and h in metres. The factor 9.81 represents the acceleration due to gravity (in metres per second squared). The power output will thus increase if flow rate or head (or both) are increased. However, the mechanical stresses resulting from an increase in head or flow rate are quite different, so turbine designs have been developed for different combinations of these two factors: for high-head systems a Pelton Wheel is recommended, whereas Francis and Kaplan turbines are recommended for progressively lower heads. These days it is common to hear of **micro-hydro systems**; these are defined as systems with an output of fewer than 100 kW; for systems ranging from that value up to 1 MW the term '**mini-hydro**' is used. The largest single hydropower turbine currently in use is rated at 784 MW, and it is one of three operating in parallel in a single hydropower station (Xiluodu) in China's enormous Three Gorges hydropower system.

By analogy, most **wind turbines** installed to date would fall in the micro- or mini- category, although this terminology tends not to be applied to them. The largest individual wind turbines are currently capable of peak outputs of 8 MW – two orders of magnitude fewer than the largest hydropower turbine. Such large turbines typically stand as much as 220 metres high from ground surface to maximum blade-tip height (140 metres to the hub), with blades up to 80 metres long. There is a strong incentive for increasing the blade length of wind turbines, as the power output is proportional to the **swept area** – that is, the circular area defined by the rotating blades – and that area is proportional to the square of blade length (remember high school maths?: $A = \pi r^2$). However, blade length can be increased only by increasing **hub height** (i.e. the height of the mast from ground surface up to the hub that holds the blades), and the larger the power output, the larger also must be the gearbox and generator housed within the **nacelle** (i.e. the housing behind the hub).

The largest wind turbines have generators with diameters of as much as 12 metres, and the engineering challenge of housing such large and weighty devices atop a slender mast is formidable. Hence, rather than install one or two very large turbines, as is done in hydropower dams, attainment of sizeable wind power outputs typically requires a large number of individual turbines clustered in close proximity, constituting a **wind farm**. At the time of writing, the largest onshore wind farm in the world has a peak capacity of around 1.3 GW (at Alta, Southern California), comprising some 490 individual turbines each rated between 1.5 and 3 MW. Future onshore wind farms as large as 5 GW are mooted, mainly in sparsely populated regions of North America. Installation of large numbers of wind turbines is complicated by the fact that wind passing through turbine blades is disturbed, usually becoming turbulent. (Turbulence is undesirable for wind turbines generally, which is why it is unwise to locate them on rooftops, cliff edges or close to large obstacles such as buildings or trees, all of which tend to give rise to turbulence.) Hence it is important to space the turbines out across a wind farm, so that they are at least 20 blade lengths apart downwind and four blade lengths apart sideways. Thus, very large wind farms have significant land requirements, and although agriculture may continue beneath the turbines, other land uses such as housing are effectively excluded, due to considerations of noise and electromagnetic interference with radio and TV signals. There is thus a marked trend, especially in densely populated parts of the world, to develop the largest wind farms offshore.

Offshore settings are in many ways very favourable for wind turbines, as wind conditions are generally more consistent than over land, and it is also easier to manipulate large structural components into place at sea, rather than fighting with the narrow and winding upland roads that tend to be found in the most favourable onshore settings. Thus, offshore wind farms are burgeoning in north-west European waters, with the world's largest offshore wind farm, the London Array, currently rated at 630 MW (comprising 175 machines each of 3.6 MW capacity). South Korea recently approved a 2.5 GW offshore wind farm, while the UK is currently planning eight offshore wind farms with capacities of around 1.2 GW each.

Whatever the size of a wind turbine, and whatever its location, its operation will be governed by three numbers: the cut-in speed, the rated output and the cut-out speed. The larger a wind turbine, the higher will be its cut-in speed, which is the minimum wind velocity required to turn the blades sufficiently to overcome the resistance to rotation arising from friction within the turbine mechanism, and thus to begin to generate usable power. Beyond this threshold, the power output of the turbine will increase proportionally with the cube of the wind velocity until it plateaus out beyond a predetermined wind speed. This plateau power output, normally displayed on the turbine nameplate, is the rated output. As wind velocity increases beyond the lowest value needed to attain the rated output, the power output will remain steady, being limited by automated adjustments of blade orientations and other aerodynamic design features. Beyond a given high wind velocity – the cut-out speed – the turbine will be automatically shut down to prevent damage. To give an example of the magnitudes of these three critical numbers: for a turbine of a size common in many onshore wind farms, with a rated output of 3 MW, the cut-in speed would likely be around 2 to 3 m/s, it would reach rated output at around 15 m/s, and it would have a cut-out speed of 25 m/s. These design features explain why many wind turbines are seen not to be working despite it being an obviously windy day.

Indeed, variability of output is one of the most frequently voiced criticisms of wind turbines. Few onshore wind farms are productive for more than 30 per cent of the time (usually expressed as a **capacity factor** of 0.3). Capacity factors for offshore wind turbines will often approach (and, in some cases, exceed) 0.4.

Maintenance considerations give rise to another peculiarity of wind farms: with a single wind farm containing many individual turbines, there are far more (albeit much smaller) moving parts than in a typical hydropower or thermal turbine plant; there is thus far greater scope for something to break down. On the other hand, temporary loss of one or two turbines will not greatly reduce the output of the wind farm as a whole, whereas loss of two turbines might entirely close a thermal or hydropower station.

As we have already seen, we are not yet close to having sufficient, cost-efficient electricity storage capacity connected to the world's power grids. Yet the very size of grids makes them resemble storage facilities to some degree. There is a clear smoothing of the erratic behaviour of individual wind farms as more of them are interconnected in a grid system, as variations in wind are rarely synchronous over wide areas. This observation has frequently been offered as a solution to the variability of wind, the argument being that, if it is not windy in, say, England, it might be windy in Spain, or vice versa. Thus, all that is needed, the argument proceeds, is sufficient long-distance interconnector cables and variability will be vanquished. There are three restraints on optimism over this argument. One is the sheer size of many weather systems that track across the continents from the oceans: it is commonly the case that a single weather system simultaneously affects sites thousands of kilometres apart. A second is the cost of long-distance interconnector cables: the latest subsea interconnector cable to be constructed in Europe (in the Irish Sea) reveals a cost of about £2.5 million per kilometre (US$7 million per mile). Finally, the more distant the interconnection, the more likely it is that the cables will need to cross territories with which diplomatic ties are weak. Thus, for instance, a plan to obtain all electricity in Europe from renewable sources envisages interconnectors to putative CSP plants extending throughout North Africa and the Middle East – through many countries with which (to put it mildly) the European powers have not enjoyed uniformly cordial relations.

In the meantime, however, at much smaller scales, many 'green power' companies sell their electricity at a premium on the grounds of being 100-per-cent renewable, relying utterly on the (largely fossil-fuelled) grid to ensure that their customers are not exposed to the chill winds of variability. It might be good business, but it relies on questionable ethics and would quickly become self-limiting were the proportion of variable sources connected to the grid to expand at the expense of disconnecting baseload and dispatchable sources.

Public opinion appears to be rather evenly split as to whether wind farms are eyesores or acceptable landscape features. Opponents of wind farms often claim that wind turbines massacre birds and bats. The bulk of the scientific evidence does not support such a sweeping generalization, though in specific circumstances (e.g. on major bird migration routes or in the breeding grounds of endangered raptors) significant avian fatalities have been recorded. On the other hand, systematic studies of the behaviour and breeding success of overwintering farmland birds in northern England revealed no negative effect from the presence of wind turbines; indeed, for small, ground-nesting birds, wind farms might actually be beneficial, as they apparently decrease the frequency of incursions by larger, predatory species. Controversy will no doubt continue to rage.

Tidal energy is also harnessed using turbine technologies. There are two principal configurations:

▶ Tidal barrages, in which dams are erected across an estuary with a high tidal range, and water is impounded at high tide to be released through hydropower-type turbines when the tide falls. Although widely touted, the only sizeable installation to date is the La Rance system in Brittany (France), which was constructed in the 1960s and comprises ten turbines in parallel with a combined peak output of 240 MW. In that system, some of the diurnal variation is smoothed by the incorporation of a degree of pumped storage. The principal drawback of tidal barrages is the very high initial capital outlay per MW installed; other concerns relate to partial loss of intertidal habitat, albeit evidence from La Rance suggests that the net ecological impact of that system has been positive. Tidal barrages are only ever likely to be worth while in areas with quite marked tidal ranges, such as the western seaboards of France and England, the Bay of Fundy in Nova Scotia, the Magellan Strait of Chile, and various estuaries in Argentina and Alaska. In contrast, areas with very modest tidal range, such as the Mediterranean, the Baltic and the Caribbean, have no real tidal barrage potential.

▶ Tidal current (or tidal stream) systems, in which devices closely resembling wind turbines are either suspended from floating buoys or fixed to the seabed in areas that experience rapid undersea currents. The first grid-connected tidal current turbine system is the 1.2-MW SeaGen device in Strangford Lough, Northern Ireland; further grid-connected systems are currently being installed off the coast of Scotland. The key challenge in developing tidal current systems relates to structural engineering: because water is 800 times denser than air, the stresses affecting tidal turbines are at least as many times greater than those that act on wind turbines. Though amenable to resolution, the necessary robustness can be achieved only at considerable cost.

Inevitably, ecological concerns have also been raised: as with concerns over bats and birds where wind turbines are concerned, many people have conjectured that sea life might be affected by tidal turbines. In reality, the blades of tidal turbines turn far more slowly than those of air turbines, and fish and sea mammals are well adjusted to spotting large objects (predators) moving towards them and taking evasive action. Certainly, extensive monitoring in Strangford Lough, including tagging and telemetric monitoring of a local seal colony, revealed no such problems in practice. So, to the degree that costs can be brought down, tidal current turbines show considerable promise.

Still rather more remote from commercial viability are the array of technologies currently being developed to harness **wave power.** Wave power is largely independent of tidal conditions, and hence the availability of potential resources is more evenly distributed around the world's oceans and large lakes. The potential prize with wave energy is high: calculations using the well-known Bernoulli equation suggest that a single metre width of deep-water waves carries 30 to 70 kW of power. However, as waves are produced by the shearing action of wind on the water surface, wave power suffers from the same variability as wind power. Together with the high density of water, the combination of high velocities and brisk changes in magnitude and azimuth of waves at the local scale ensures that the design, installation and operation of **wave energy conversion devices** are beset with engineering challenges. Full-scale sea trials of a range of wave technologies are under test at the European Marine Energy Centre (EMEC), based in Orkney, Scotland. EMEC is eminently well located, as Scotland is estimated to host 25 per cent of Europe's tidal energy resources and 10 per cent of its wave energy resources.

Wave motion can be conceived as comprising three elements:

1 heave: the vertical oscillation at any one point (e.g. a tethered boat bobbing up and down)

2 surge: the lateral displacement, which good surfers exploit so gracefully

3 pitch: the irregular seesaw action, which is often the prelude to seasickness.

Wave energy conversion devices are typically designed to harness one of these three motions. Several harness heave by trapping oscillating water columns within pipes, where they compress air as they rise, driving it through an air turbine, and suck air back from above as they fall. The Wells Turbine is perfectly suited to this duty, as it is designed to rotate in the same direction irrespective of which side the air current comes from. Other devices harness surge, such as the near-shore OYSTER device (proprietary technology of Aquamarine Power, tested at EMEC), which is essentially a spring-loaded hinged flap fastened to the seabed; lateral movement of the flap compresses water in a pipeline connected to shore, where it drives a conventional hydropower turbine. Pitch and heave are both exploited by the Pelamis Wave Power's 'sea-snake' device, which is tethered to the seabed in deep water. Each Pelamis device comprises five rigid, buoyant, tube-like members, connected in a single long chain by flexible joints. As waves pass the Pelamis, squeezing of the joints operates hydraulic cylinders that pump fluid into high-pressure accumulators, which then produce a constant current on board that is simply passed ashore via standard subsea power cables.

Turbines are also used to generate electricity from solar energy, by means of large so-called concentrated solar power (CSP) plants, in which parabolic mirrors are used to reflect the Sun's rays on to a water tank mounted on a central tower, producing steam that then drives a turbine in accordance with the Rankine Cycle. By storing thermal energy in salt that is melted by day and then quenched at night to produce steam, CSP technology can produce power for much of the time. It is only really viable in regions that reliably receive intense solar radiation for most of the year (e.g. circum-Mediterranean desert regions and the south-west United States). However, CSP requires a significant water supply, not only for steam-raising (much of which can be recycled), but also to operate the condensers, which must be used to cool the turbine exhaust gas if the system is to operate at an acceptable efficiency. Deserts are short of water, so this is an important constraint on CSP uptake. The long distances from many ideal desert sites to large demand centres, though amenable to technical resolution using high-voltage DC power transmission, can also be an economic obstacle.

No water is needed to operate solar photovoltaic (PV) technology, which is the only widespread electricity source yet available which does not depend on some variant of turbine technology. Rather, solar PV works by exploiting a difference in charge between two juxtaposed semiconductor materials (usually silicon), one of which is 'doped' with positive ions, the other with negative ions. The flow of electrons across the interface between the two layers (called a '**p-n junction**') is harnessed using metallic contacts on either side of the panel, directly producing electricity. The prize for efficiently harnessing solar radiation for power production is extremely enticing, as the average net solar radiation available at the Earth's surface dwarfs current world energy consumption by a factor of more than 10,000. Much of this incoming radiation is consumed in driving winds, producing rain, and in photosynthesis. Nevertheless, there is still more than enough solar energy potentially available to more than meet even the most extravagant demands of present and future human populations. So why do we not ditch all other technologies and pin all our hopes on solar PV? There are seven principal reasons:

1 **Temporal variability:** solar radiation varies, not only diurnally but also seasonally. For instance, while a square metre of the ground surface in Northern Europe might receive 14 to 18 MJ of solar radiation over a long day of unbroken sunshine in July, it will not receive more than 4 MJ (and, in more northerly parts, not even 2 MJ) on a short January day, however fair the weather.

2 **Spatial variability:** where a solar panel rated for a peak output of 1 kW might give an annual yield of 6.5 GJ in Southern California, the same panel would yield only 3 GJ in central Germany and only 2.5 GJ in Scotland. Given that the capital cost of the panel is roughly the same in all three places, the '**levellized cost**' of solar PV electricity (i.e. with all costs spread out over the full lifetime of the panel) will vary markedly from one latitude to another.

3 **Efficiency:** the most efficient solar PV panels currently available on the market cannot convert more than 23 per cent of the incoming solar radiation into electricity. Novel designs currently in research laboratories can already achieve almost twice this figure, though they remain too pricey for early commercialization.

4 **Power density:** even the most efficient solar PV array in the world will still be limited by the absolute amount of solar radiation arriving on a given area, and this will never match the power densities achievable with fossil fuels and nuclear power. This means, for instance, that there will never come a day when on-board PV is the predominant power source for shipping.

5 **Mode of energy delivery:** PV produces DC power and significant losses are involved in converting it to AC, let alone to other forms of energy altogether, such as liquid or gaseous transport fuels.

6 **Durability:** exposed to the elements as they are, solar PV cells cannot be expected to last for ever. There are a growing number of examples in which the payback period for PV arrays (i.e. the time required for revenues from solar power production to exceed the initial installation cost) fails to exceed the functional life of the unit, due to weather damage.

7 **Cost structure:** long dismissed as exorbitantly expensive, solar PV has been on a steep trajectory towards competitive levellized costs over the last decade, thanks to innovations in materials and systems designs. Indeed, solar PV electricity may

even reach parity with fossil-fuel-derived electricity in the next decade. However, as nearly all the costs with solar PV are incurred up front (a feature it shares with wind, wave, tidal and geothermal power), it is difficult to ensure its uptake when the people incurring the capital costs (e.g. house builders) are not the same people who will benefit from the long-term cost of power (house buyers or tenants).

But if solar PV is not a panacea, it certainly has an increasingly important contribution to make. It is already the technology of choice where relatively modest amounts of power are required in remote locations, such as powering rural road signs, bus shelter lights, fuel station forecourt lights and small weather stations. In urban areas, PV is now commonplace on the roofs of many houses, thanks to generous publicly funded subsidy schemes. However, unless you have a very large roof and peculiarly low levels of power usage, it is highly unlikely to meet a large proportion of your daily electricity needs. Furthermore, times of peak demand for domestic power tend to be around the times of breakfast and evening meal, when sunlight is naturally weak or absent. The obvious solution to this problem is to store power from the middle of the day to use at other times. However, as battery storage is very expensive, most PV users bypass this problem by maintaining a connection to the national power grid, which they then use as if it were a giant cheap battery. Indeed, in several jurisdictions domestic PV users are paid handsomely for the power they sell to the grid, often at rates that privilege solar above all other forms of energy production. This raises the broader issue of how best to operate power grids in future, to which we now turn.

7

Power to the people: energy grids in transition

'Coming together is a beginning; keeping together is progress; working together is success.'

Winston Churchill

▶ National energy grids in a changing climate

For about a century now, the inhabitants of the industrialized countries of the Global North have been used to the concept of receiving energy in two forms – as electricity and as gas for use as a heating fuel – from large-scale infrastructure networks organized at national scale. Without exception, those countries of the Global South who do not yet benefit from full national coverage by electrical grid systems are vigorously pursuing them as an absolute priority. Politicians in sub-Saharan Africa fall over one another to promise expansions of grid connectivity: in many East African countries, for instance, less than 15 per cent of the population is currently grid-connected, and the ambition is to achieve full connectivity in years rather than decades. This is one of the key factors driving a projected increase in global energy consumption, from around 530 exajoules (EJ) in 2014 to around 700 EJ in 2030 – a 32 per cent increase over 16 years. There is clearly a huge appetite in developing countries for the benefits that large-scale interconnectivity brings in terms of the security of energy supply for individual citizens.

Yet, curiously, in those very countries that have benefited from national power grids for longest (e.g. the UK and Germany), the move to increase localized power generation using renewable technologies has led some enthusiasts to suggest that the very notion of a national grid is outmoded and ought to be abandoned. Although this is not (yet) official policy in any such country, the tensions arising from increased uptake of renewables subject to rapidly varying outputs (especially wind and, in Germany, solar PV) is leading to a *de facto* situation in which long-established grid systems are increasingly struggling to keep up with creeping demand for fundamental retasking: power grids that were designed on the assumption that a small number of large power stations with controllable outputs would feed distant urban populations are now being asked to cope with a large number of (very) small power production facilities of weather-dependent output scattered widely across the national territory.

In Chapter 4 we learned how the energy industry has traditionally matched supply and demand by organizing constant baseload power output, and then scheduling power stations with variable output capabilities to dispatch (or stand down) given wattages over certain periods. This is fine as long as the power stations connected to the grid have baseload or dispatchable capabilities predictable a few days in advance: currently, this means conventional power stations and, among the renewables, hydropower, geothermal, biomass and tidal. However, for those renewable sources that cannot be reliably predicted more than a few hours in advance (wind, solar and wave, for instance), real grid management challenges can arise once the proportion of total grid-connected capacity rises into double-figure percentages. It is these challenges – rather than any more sinister motives or character flaws – that have historically led grid managers to deprioritize wind and solar. Coupled with their traditionally high installation costs, there was little likelihood of the open market favouring their widespread connection to national power grids. However, as concerns over climate change have prompted governments to prioritize low-carbon power sources, and as new industries have matured and achieved economies of scale, many European countries now have installed capacities of wind/solar exceeding 20 per cent of the total grid-connected power production fleet. Above that level of 'grid penetration', it becomes increasingly difficult to maintain a stable grid frequency, which is critical to the operation of common end-user appliances. Such challenges have recently prompted remarkable statements from those responsible for managing large grids, such as the CEO of the UK's National Grid company, Steve Holliday, who commented in 2011 that 'The grid's going to be a very different system in 2020, 2030. We keep thinking [that] we want [the grid] to be there and provide power when we need it ... We're going to have to change our own behaviour and consume it when it's available, and available cheaply ...' (BBC Radio 4, *Today*, 1 March 2011). How serious is this suggestion? Could any elected government survive the abolition of electricity-on-demand? What does all of this mean in practical terms?

Grid-locked: how Germany's low-carbon transition led to increased CO_2 emissions

Germany can be rightly proud of its *Energiewende*, that is, its enthusiastic transition towards deployment of renewable energy technologies, which has seen it become a world leader in per-capita installed capacities for wind and solar power. This has been achieved by offering impressive public subsidies for these technologies, exceeding retail costs for fossil-fuelled electricity by factors of around 2 (for wind) and 10 (for PV). But this has so privileged these sources of power on the grid that many highly efficient combined-cycle gas turbine (CCGT) plants have become uneconomic and closed down. Any hope of redeeming the economics of German CCGT plants, which currently rely on imports from Russia, were dashed when, in response to vociferous campaigning, the German government banned shale gas exploration. Around the same time, the government also responded to the Fukushima incident in Japan by announcing that it would accelerate the closure of all remaining nuclear power plants.

However, since the burgeoning wind and solar installations could not provide baseload or dispatchable duty, it soon became apparent that the grid was heading for sporadic, lengthy power outages. Evasive action was taken by the German government, which has turned to its cheap indigenous lignite resources to support up to 10 GW of new lignite-fired power stations, the first of which (>2.2 GW) was recently commissioned near Cologne. The collapse of the value of carbon credits on the EU emissions trading system has apparently made German lignite irresistibly cheap. The problem is that lignite – a form of coal akin to peat – has a much lower calorific value than bituminous coal and significantly higher carbon emissions: it is, in fact, the worst of all the fossil fuels from the perspective of climate change mitigation, with more than twice the CO_2 emissions of the gas it has effectively replaced. In these circumstances, the only hope of avoiding a rise in carbon emissions would be to rapidly fit carbon capture and storage (CCS) facilities to these new lignite-fired power stations. Yet the only pilot-scale CCS project yet attempted by the German authorities was cancelled when the government failed to transpose EU CCS law into national legislation on a reasonable timescale, with legislators backing down in the face of vociferous opposition by pressure groups. Controversy focused largely around subsurface injection of CO_2, though the objectors' arguments betrayed their utter lack of hydro-geological understanding. This concatenation of unintended consequences has already begun to manifest itself in a rise in Germany's total annual carbon emissions: having declined steadily since the 1990s to reach a minimum of 917 million tonnes (Mt) in 2011, they subsequently rose to 931 Mt in 2012 and 951 Mt in 2013. The moral of the story is: good intentions alone will not balance a national power grid.

▶ Power transmission and distribution

Most of us are conscious of the voltage that applies to our domestic electricity (i.e. 110 volts (V) in the Americas and much of mainland Europe, 240 V in the UK and many Commonwealth countries). However, the voltages at which electricity is released by power stations is typically a hundred times greater – for instance, around 25,000 V (25 kV). It might seem that all that would be needed would be to step the voltage down to the consumer levels and deliver it: but this is fundamentally mistaken. Manipulation of Ohm's Law leads to the crucial insight that the loss of power along a cable is inversely proportional to the voltage. Hence, to minimize power losses in delivering electricity from a generator to a consumer, the transmission cables should be operated at as high a voltage as the materials composing the cable will stand. It is for this reason that long-distance power lines are operated at very high voltages: in the UK, for example, voltages of 132 kV and 400 kV are used for long-distance transmission. Nearer to end-users, substations are used to step the voltage down again for localized distribution, first to 66 kV (at which voltage heavy industrial users are commonly served), then to 33 kV and 11 kV (serving smaller industrial users). At the end of the distribution system, 11 kV power is converted to 400 V in small distribution transformers, whence cables serve domestic properties.

Until recently, national-scale power grids have been configured to operate solely using **alternating current (AC)**. This was in many ways an obvious choice, as most large power-generation plants produce current by rotating magnets between coils of wire – a process that results in the alternations of positive and negative flows of electrons that constitute alternating current. AC is also highly versatile when it comes to delivering power to diverse end-users (industrial, domestic) at different voltages. The only real drawback with AC in large-scale power grids arises from some important technicalities of cable performance, which mean that individual cables cannot be longer than a few tens of kilometres. To get over this problem, it is possible to handle long-distance transmission by using **direct current (DC)**, which comprises a steady, unipolar flow of electrons. There is no length restriction on DC cables, which makes them extremely useful for very long-distance transmission, especially via subsea cables, where installation and operation of conventional AC relay stations are both difficult and costly. Furthermore, DC cables are less prone to transmission losses than AC cables. High-voltage DC (HVDC) cables are thus increasingly the technology of choice for long-distance transmission, such as subsea international connectors. In such scenarios, power will be generated in AC form, then rectified to DC for transmission, before being converted to AC for final distribution.

This is by no means the only use of DC, as all batteries and some small power-generation devices (e.g. dynamos and solar panels) yield DC power, and many low-voltage modern appliances consume DC power – hence the proliferation of small, heat-emitting chargers (which, technically, are AC to DC rectifiers) for laptops, mobile phones, MP3 players, gaming consoles and so on. Lamps made up of clusters of light-emitting diodes (LEDs) also require DC power. As rectifying AC to DC and converting DC to AC incurs inevitable losses of energy – as we would expect from the Second Law of Thermodynamics – it is best to keep such transformations to a minimum. Yet so tight is the grip of technology lock-in that, even where houseowners are generating DC power on their roofs using solar PV, to connect it to the domestic wiring circuit it is routinely converted to AC in an inverter box – losing as much as 25 per cent of the power in the process – before being rectified back to DC (losing a further 20 per cent or more) for use in modern appliances (which already account for between 15 and 30 per cent of a typical household's electricity consumption). A smarter use of domestic solar PV would be to set up a separate DC circuit in the house and use this to support all

DC-using devices. DC micro-grids are already emerging for data centres, in which large numbers of computers are running constantly: at present, large amounts of money are spent on cooling these buildings, simply to remove the heat produced by AC to DC rectification.

Domestic PV is but one example of distributed energy-generation technologies increasingly being installed at the low-voltage extremities of distribution systems; others include wind turbines and **micro-CHP** (i.e. domestic gas boilers that are designed to generate some power as well as drive central heating systems). To the uninitiated, it may seem like a simple matter to connect these generators up to the nearest power cable and sell power back to the grid. However, given that existing grids were designed decades ago to transmit power in one direction, this is far easier said than done. Indeed, it is not much less challenging than proposing to send water from your sink up the tap and back to the waterworks. Simply connecting into a 400-V neighbourhood distribution line can serve only very local needs – and even then, without careful design and management, variable inputs of power can disturb the supply received by neighbouring properties that share a 'point of common coupling' to the wider distribution network. It would be best to connect the local generators into the 11 kV distribution network – though this invariably requires the laying of new cables and investment in transformers and switch gear, none of which is simple or cheap. The question then immediately arises: who should pay for such grid adaptations? Or are there ways we can adapt the existing grid to work better with the new distributed power sources, to minimize the expense of substantially rewiring much of the country? Can we make our electrical power systems work *smarter* rather than harder?

▶ The promise of smart grids

Such questions have prompted the development of a whole new field of endeavour in electrical and electronic engineering: **smart grids.** Various definitions of 'smart grid' have been offered, of which the following is fairly representative: 'A Smart Grid, as part of an electricity power system, can intelligently integrate the actions of all users connected to it – generators, consumers and those that do both – in order to efficiently deliver sustainable, economic and secure electricity supplies' (Electricity Networks Strategy Group, UK, 2009). It has to be said that, with the exception of small pilot systems, the concept of a smart grid still remains somewhat aspirational. But it is an aspiration with huge international momentum. So what sorts of technologies and operating procedures can we expect to deliver on the smart grid vision?

Central to the smartening-up of grid systems is data. When grids had to operate as one-way systems, from power station to power-point, many operational decisions could be safely based on broad assumptions about system properties and responses. When the lower-voltage reaches of distribution systems can dynamically switch between import and export of power, broad assumptions are no longer adequate. If analyses of system capacities are based on extrapolation from the former dispensation, there is a distinct risk that distributed generation might be either excluded from the grid unnecessarily or else accepted inappropriately, thus increasing the frequency of local grid faults and resultant power outages. For instance, every power line has its maximum carrying capacity, beyond which overheating may occur. In the past, it has been common practice to simply assume static, 'one size fits all' thermal ratings for each category of line, and to manage currents/voltages accordingly.

A very concrete example relates to wind energy. By definition, wind farm outputs are at their highest when winds are fairly strong. Yet these are the very times at which wind chill leads to much greater cooling of overhead power lines. It has been shown that the capacity of overhead lines can more than double during windy weather; yet if traditional static thermal ratings were used to estimate system capacity, wind power would often be unnecessarily rejected from the grid. New technologies that directly sense the true rating in real time have the potential to expand the carrying capacity of elements of the grid without any rewiring.

Real-time data on power use within homes and businesses also have the potential to be transformative. Assumptions on power use in homes, for instance, are often based on one-off surveys from many years ago. Yet, as we all know, magnitudes and modalities of power use have fundamentally changed over the last decade, thanks to the boom in consumer electronics. Not only would more accurate profiles of energy use be helpful to grid managers; they could empower consumers to become far more savvy about how and when they use power, allowing them to take advantage of those periods when electricity prices drop due to a surplus of supply over demand. Collection of such information is the purpose of **smart meters**, which will share information between suppliers and users instantaneously.

Small-scale and distributed **electrical energy storage** is another technology expected to play a key role in future smart grids. One possibility immediately available would be to enter into agreements with householders to use their refrigerators/freezers as dynamic energy stores, given that such devices can fulfil their primary food storage role even though their compressors run only intermittently. Equally, there is no harm done if a freezer gets extra cold for a short period by running its compressor a little longer than usual. If agreements between householders and power suppliers can be reached, smart domestic appliances could be used to help balance local grid loads. Similar possibilities would arise in the event of widespread uptake of battery-powered electric vehicles that are grid-connected when parked.

It is clear that smarter grid management will only really be achieved if we get far better at energy **demand minimization.** Behavioural change has a role to play in this, and smart meters could certainly help to encourage this. Innovations in power electronics are already making common domestic appliances far more efficient than hitherto, and can compensate for neglectful owners by down-powering automatically when left unused. Substitution of LED lights for traditional AC light fittings – and for the unpopular slow-start fluorescent tube low-energy light bulbs that have replaced them by edict in recent years – can make a big difference to lighting throughout the built environment, as can automated light-dimming during periods of human inactivity. Perhaps the greatest single contribution to demand minimization is the least glamorous: enhancing insulation of buildings to eliminate heat loss, thus cutting the need for the single greatest domestic use of energy, and the greatest use of energy overall: space heating and cooling.

8

Secure, green and cheap energy: can we *really* have it all?

'Human progress is neither automatic nor inevitable. Without persistent effort time itself becomes an ally of the insurgent and primitive forces of irrational emotionalism and social destruction. This is no time for apathy or complacency. This is a time for vigorous and positive action.'

Rev. Dr Martin Luther King, Jr

▶ Energetic arguments

If you, like me, feel compelled to follow the great energy debates of today, in the popular press and on protestors' placards, perhaps you have also noticed that most of the arguments spectacularly miss the point. Certainly, in the UK and most of Europe, you could be forgiven for thinking that electricity is the predominant form of energy consumption. Yet, as we have repeatedly noted in this book, that distinction really goes to the far less glamorous heat, with transport fuels in second place. Really, at only 20 per cent of total energy use, electricity is a bit of a sideshow, which we could safely ignore for the time being if only we were really getting on and making progress on heat and transport. Yet many activists are apparently oblivious to the irony of driving gas-guzzling 'vintage' camper vans (festooned with green slogans) to protest camps, where they warm themselves and cook using propane/butane camping stoves. (Just for the record, there is no known 'green' source of propane or butane.)

And make no mistake, protests against *all* energy technologies have become a permanent feature of civil society in the early twenty-first century. The list of energy developments that have been the object of vociferous protests in Europe since 2000 is daunting: wind turbines and wind farms; tidal barrages and submerged turbines; the covering of farmland with solar panels; gas pipelines; overhead power lines; nuclear power stations; anaerobic digesters; coal mines; coal-fired power stations; gas-fired power stations; hydropower developments; biomass power plants; CCS pilot plants; shale gas wells and 'fracking'. It would be interesting to know how many doughty campaigners have protested against *all* of these at one time or another.

As a superannuated protester myself (it was nuclear weapons, apartheid and neo-fascist political parties we marched against in my day ...), I can testify that it is not only easy but positively pleasurable to say what you are *against*. It gives you a warm glow of moral authority, enhanced by the tremendous sense of fellowship with your co-agitators. Saying what you are in favour of is, in my experience, far more difficult: it does not lend itself so readily to sloganizing, and angry words come less readily to the tongue.

Yet behind all the *anti-* protests lie some positive aspirations. Energy protesters typically cite one or more of the following four grounds for protest:

1 visual or auditory intrusion (which is often a proxy for 'impact on property prices')

2 localized ecological impacts

3 climate change impacts

4 costs of energy to bill payers.

If we take the reverse of these four grounds, two positive desires emerge: there is a consensus that people want their energy to be 'green' (i.e. environmentally sustainable) and cheap.

▶ The unrealized imperative: energy security

Energy systems have a third desirable attribute, which has largely been absent from recent debates and certainly from recent protests: **energy security**. This has been formally defined by the International Energy Agency as follows: 'Energy security means the uninterrupted availability of energy sources at an affordable price.'

For the majority of humankind, resident in the Global South, energy security thus defined remains but a cherished dream, the realization of which is the foremost priority in most democratic jurisdictions. For the privileged populations of the Global North, on the other hand, energy security has been taken for granted for generations. It is tacitly assumed to be a 'given', as natural as the availability of fresh air. Yet this is a dangerously fragile assumption. In urbanized societies, energy security is assured only by the maintenance of complex chains of fuel supply and energy grids (gas and electricity). We have already explored some of the vulnerabilities currently attending these in Chapters 3, 4 and 7. Yet, in all the energy protests of Europe, there seems to be no recognition at all of the importance of energy security. I strongly suspect this will remain the case until complacency is shattered by the onset of frequent and sustained energy '**outages**' or **blackouts.**

Is it not ludicrously alarmist to suggest that such a turn of events might soon come to pass? We have already seen (in Chapter 7) how close Germany has come to being unable to maintain baseload and dispatchable capabilities on its grid, as capitulation to a string of single-issue energy campaigns left the country dangerously exposed to the inherent variability of wind and solar power. Germany is by no means alone in this predicament. On 19 February 2013 the outgoing head of Ofgem (the UK's energy industry regulator) warned that recent closures of nuclear and coal-fired power stations in the UK have removed the traditional 20-per-cent reserve of generating capacity over peak demand that has been maintained since the mid-twentieth century, so that by 2016 'the reserve margin of generation [will] fall from about 14 per cent [in 2013] to less than 5 per cent. That is uncomfortably tight ...' Following privatization of its energy industry in the 1990s, there has been no obligation on UK power companies to invest in new generation plant, and they have therefore tended to opt for paying dividends to shareholders. Consequently, the UK has not built a single sizeable power station with baseload/dispatchable capability since 1990; the only recent investments have been in wind farms, which boast neither capability. With new coal-fired stations effectively ruled out under UK climate change policies, the only technology that can be built to sufficient scale in time to avoid blackouts is gas-fired turbines. The outgoing head of Ofgem goes on to argue that '... by 2020, 60 per cent to 70 per cent of our (power) generation may have to come from gas to fill the gap. That's up from about 30 per cent today ...' It remains to be seen whether the required CCGT plants will indeed be constructed in time. The energy security story doesn't even end there. The UK's production of conventional natural gas has been in steep decline for many years now, so, unless unconventional sources are developed, dependence on imports will continue to grow sharply. Then bear in mind that two-thirds of the gas consumed in the UK is used in houses, with 82 per cent of households relying on gas for their heating, hot water and cooking, and the scene is set for a perfect storm. Meanwhile, UK protesters carry placards demanding 'Less gas, more wind'. Show me the first baseload or dispatchable wind resource in the world (God will need to be on board for that one), or the first house which is heated rather than chilled by the fickle winds of winter, and maybe I'll be persuaded. In the meantime, it would appear that more wind necessarily demands more gas also.

▶ The environmental trump card

As an environmental engineer who spent the first decade of his career working on the impacts of climate change on water resources, I consider the debate over climate change to be settled. I will not waste space here on the special pleading of supposedly sincere deniers of climate change whose CVs always betray strong vested interests in 'business as usual'. Human-induced climate change is real and is increasingly posing existential threats to humans and ecosystems, especially in the

Global South. We cannot afford to continue this game of atmospheric roulette.

Yet, whenever we attempt to address the climate change imperative by proposing renewable energy systems, nuclear generation, or fitting CCS to all fossil-fuelled power plants, up pop the 'environmental' protesters. Few of them are compromised by the ulterior motives of the most prominent climate change deniers, yet still it is difficult not to feel exasperated that extreme application of the precautionary principle, on even the most unexceptional of sites, can always be deemed to trump the global environmental imperative of combating climate change. Yet, on what basis can we ever decide that at least some impairment of ecosystems in numerous localities is a lesser evil than permitting climate change to proceed unchallenged? There is no one human authority that can adjudicate; separation of powers reigns supreme. Just as a coherent energy policy needs to be much more than a crude amalgamation of capitulations to single-issue protests, so a coherent environmental policy needs to take a holistic view of ecological impacts across the entire temporal and geographical life cycle of energy resources. For instance, nuclear energy is undoubtedly low-carbon, yet former uranium mines have given rise to severe cases of acidic mine drainage in some localities, and the most potent radioactive wastes require isolation from the biosphere for many millennia. How do we assess all of the benefits and impacts at once, and thus reach a rational decision on the optimum balance?

▶ Economic paradigms for energy management

The political consensus for the last few decades has been to leave such decision-making to the markets. Privatization has become a powerful orthodoxy in the energy sector that few dare to challenge. Yet, judged against the challenges of reconciling energy security and environmental exigencies with affordability, the market has clearly failed to deliver. At the same time as nonchalant approaches to balancing energy grids are beginning to backfire, fuel poverty and atmospheric CO_2 concentrations are both mounting. To give privatized energy its due, it has delivered energy at bargain-basement prices for a couple of decades. Yet it has achieved this only by neglecting to absorb the true environmental costs of unabated greenhouse gas emissions. As that approach becomes morally redundant, all of the options converge on one conclusion: we are going to have to pay more for our energy if it is to be secure and green.

Detractors of wind, solar and nuclear energy routinely condemn them for being far too expensive compared to conventional energy sources. If we look at estimates of levellized costs of energy delivered (e.g. in US dollars or pounds sterling per kWh) from a worldwide range of sources, a far more interesting picture emerges. Consider the two diagrams in Figure 8.1.

The real marker in these is onshore wind: unlike many of the other renewables listed, this is already a mature technology, for which substantial reductions in cost are

Current market costs without CCS

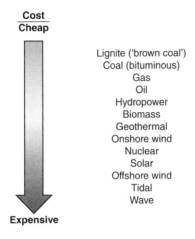

Fairer comparison <u>with</u> CCS

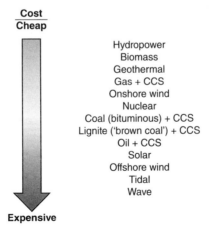

▲ Figure 8.1 Relative levellized costs of currently available energy technologies.

no longer anticipated. The shift in the relative position of onshore wind between the first and second diagrams is therefore instructive. The conclusion is clear: if fossil fuels are forced to internalize the costs of climate change in the most sincere manner possible – that is, by proceeding only if they are fitted with carbon capture and storage (CCS) facilities – then many renewables and nuclear are already as cheap, or even slightly cheaper. Of course, this does not solve the intricacies of which energy sources are technically suitable for delivering heat and transport, nor of which are capable of producing electricity on demand. But in a truly levellized playing field for energy technologies, it is already clear that we need to bring renewables, nuclear and abated fossil fuels together in an optimum mix to ensure energy security: they all have important roles to play, and they all cost much the same when we truly compare like with like.

The economic corollary is ineluctable: if we really want our energy to be secure *and* green, it simply *cannot* remain as cheap as it has been in the long era of unabated fossil fuel use. This point has already been tacitly accepted by many governments. For instance, the UK government recently sought to stimulate nuclear new-build by guaranteeing a minimum price for the power produced that amounts to more than double the current price of fossil-fuelled electricity. Yet, at the very same time, politicians were railing against the privatized energy companies, claiming that present energy costs are an outrage. Whatever the rights and wrongs of companies' transactions, any politician who tells you that we can have secure *and* green energy without a substantial price rise (an approximate doubling) is either ignorant or mendacious.

Given that a serious response to climate change renders such a price rise necessary, how do we deal with the wider economic fallout? Energy costs for industrial production are a clear concern. It is not easy to envisage a free-market solution to that challenge in the absence of a binding global agreement on an upper limit for the carbon intensity of national economies (in Mt of CO_2 per unit GDP, for instance).

Of even more concern is the impact on the poorest households: those in **fuel poverty.** In developing countries, absolute poverty severely limits the ability of people to participate in the energy economy. Yet even in the Global North, fuel poverty is a growing issue. Although definitions vary, in UK terminology a household is considered to be in fuel poverty when it has to spend more than 10 per cent of its income to maintain an adequate standard of warmth. (Here, at last, is one governmental message on energy that zooms straight in on the predominant energy use category – heat.) Clearly, multiple factors can lead a household into fuel poverty: the absolute cost of fuel; poor house insulation; the increase in single-person/ small households; and low household income. Hence the metric should be used with care: because her palaces are large and draughty, Her Majesty the Queen is officially in fuel poverty! In this, she shares (statistically at least) the fate of about a quarter of the UK population – most of whom live not in palaces but in poorly insulated, rented properties. In rural areas, where most households cannot access comparatively cheap mains gas, the incidence of fuel poverty rises above 50 per cent. If it is difficult to foresee a free-market solution to rising industrial energy costs, it is impossible to see one for fuel poverty. If we really want energy that is secure, green and affordable for all, then we need to reinvent our energy policies with solidarity as a guiding principle.

So, then, it is clear that we *can* have **secure and cheap** energy: that is exactly what fossil fuels have delivered for more than a century. It is also clear that we *could* also have **secure and green** energy, provided we do our best to minimize energy demand, and back up the inherent variability of mainstream renewables with decarbonized fossil fuels (using CCS) and nuclear power (especially using thorium and, perhaps, eventually, cold fusion). But if current prices for energy produced from unabated fossil fuels are to be the measure of cheapness, then it is clear that we *cannot* have secure, green *and* cheap energy. I unflinchingly advocate secure and green energy, but we have to grow up and accept that this will, in round terms, cost at least twice what energy has cost us hitherto. (If the US shale gas revolution is maintained and repeated elsewhere, the cost gap might widen further, though shale gas with CCS would still be more expensive than current bulk energy costs.)

We *can* solve the energy trilemma; we just need the collective will to accept the costs, and ensure that we do not abandon those who cannot afford an increase in costs. The stakes for ecosystems and human civilization could scarcely be higher. So let's forget the facile slogans and the single-issue campaigning, and work together to identify the optimal blend of technologies to deliver heat, transport and electricity securely and sustainably.

This **100 ideas** section gives ways you can explore the subject in more depth. It's much more than just the usual reading list.

Twenty pioneers of energy science and engineering and their contributions

It is symptomatic of the waste of talent that male domination has entailed that only one of the following 20 pioneers is female ... At least a fair few are genuine cases of people from very humble backgrounds achieving great things in the face of prejudice.

1 **André-Marie Ampère** (1775–1836). Ampère's theorem established the principles of electromagnetism – which remains the basis of most modern electricity-generating devices. His inventions include the solenoid, the galvanometer, the electric telegraph and, of course, the electromagnet. The unit of electrical current (the ampere, usually abbreviated to amp) now bears his name.

2 **Antoine Henri Becquerel** (1852–1908). A Parisian engineer and physicist, Becquerel discovered natural radioactive decay and went on to identify several of the most practically useful radioelements, notably thorium, polonium and radium. The SI derived unit for radioactivity is the becquerel (Bq), which is the radioactivity produced by disintegration of one atomic nucleus per second.

3 **John Buddle** (1773–1843). A major innovator in coal mining, he developed more effective ventilation systems, lining technologies for shafts, safer methods of lighting (he assisted Sir Humphrey Davy in developing his famous safety lamp) and steam-assisted coal haulage technologies.

4 **Anders Celsius** (1701–44). A distinguished astronomer, he is most remembered for pioneering the 100-point scale of temperature running from the freezing point to the boiling point of water. Anders' original version of the scale labelled the freezing point as 100 and the boiling point as 0; the order was reversed to today's familiar scale only after his death. A degree Celsius is defined as the absolute temperature (in Kelvin) −273.15.

5 **Nicolas Léonard Sadi Carnot** (1796–1832). Sadi Carnot, as he was always known, only ever published one document, but it turned out to be a fundamental contribution to thermodynamics, as it established the theoretical maximum efficiency of any heat-driven engine. This is a useful yardstick, providing an upper bound to which any such device could aspire, in the absence of losses due to friction, conductive loss of heat, etc.

6 **Rudolf J.E. Clausius** (1822–1888). One of the principal founding fathers of thermodynamics, he recognized an inconsistency between the Carnot Cycle (as then understood) and the principle of conservation of energy (i.e. the First Law of Thermodynamics). So he devised the Second Law of Thermodynamics, introducing the concept of entropy in the process. He also independently derived a version of the Ideal Gas Law from first principles.

7 **Benoît Paul Emile Clapeyron** (1799–1864). His work on 'the motive power of heat' clarified and advanced the work of Carnot and pulled together the earlier work of Boyle and Charles to produce the Ideal Gas Law. He also built on Clausius' work to produce the first rigorous explanation of the processes that occur during phase changes.

8 **Marie Curie** (née Sklodowska) (1867–1934). She made many of the most fundamental discoveries concerning radioactivity, which later paved the way for development of nuclear energy.

9 **John Dalton** (1766–1844). A meticulous and rigorous meteorologist, much of Dalton's work provided the scientific underpinnings for accurate observations of weather and climate upon which present-day characterization of wind and hydropower energy resources, and assessments of climate change induced by fossil fuel use, effectively build. Of even greater significance was Dalton's work as the principal originator of atomic theory, which underpins all later nuclear energy developments.

10 **Rudolf Diesel** (1858–1913). An early refrigeration engineer who experimented widely with heat engines, including a very early solar-powered device, he is best remembered for the highly efficient variant of the internal combustion engine that he developed. This, together with the refined variant of petroleum on which it runs, continues to bear his name and remains the staple transport energy to this day.

11 **Thomas Edison** (1847–1931). Like many great industrialists, Edison's genius lay not so much in origination but in honing new technologies so that they became commercially viable. Thus, his first hydroelectric power plant was not the first ever (that was built by Sir William Armstrong in the UK) but it paved the way for many more. He was also narrowly beaten to the invention of the electric light bulb by another Briton (Sir Joseph Wilson Swan), but after a brief legal wrangle the two inventors pooled their efforts and the EdiSwan company delivered electric lighting to the masses for the first time.

12 **Albert Einstein** (1879–1955). Widely regarded as the most outstanding scientist of the twentieth century, Einstein introduced concept of special relativity (including the famous equation $E = mc^2$ – see Chapter 2 – the cornerstone of nuclear energy production) and general relativity. He later delivered the first rigorous explanation of the photoelectric effect – the keystone of modern solar PV technology.

13 **Michael Faraday** (1791–1867). He is best remembered for his discovery of electromagnetic induction, the breakthrough that transformed the insights of Ampère and others into one of the foundational technologies of electrical power generation. Faraday effectively invented the first electrical transformer and the first DC electrical generator. Subsequently, Faraday worked on electrolysis, establishing – together with Davy – the principles of electrochemistry: the science that underpins all modern batteries and fuel cells.

14 **James Prescott Joule** (1818–89). A brewer by trade, James Joule had to pursue science as a hobby. He discovered the First Law of Thermodynamics – that is, the principle of conservation of energy. Working with Kelvin, he helped to establish the absolute temperature scale. He also established the relationship between electrical

conduction through a wire and the release of heat: Joule's First Law. Among many other things, this law underpins modern grid-management approaches. The SI unit of energy, the joule, is named after him.

15 **Charles Hesterman Merz** (1874–1940). A pioneering electrical engineer, he is best known for mastering three-phase, high-voltage AC power distribution – an experience that paved the way for the establishment of the UK National Grid, which subsequently provided a template for similar grids in numerous countries worldwide.

16 **Georg Simon Ohm** (1789–1854). He discovered that the current that flows through a conductive wire is directly proportional to the voltage to which it is subjected. This observation was codified as the well-known Ohm's Law. The unit of electrical resistance, the ohm, is named after him.

17 **Charles Algernon Parsons** (1854–1931). He developed the first efficient steam turbine and delivered it to the electricity and naval propulsion industries (heralded by the *Turbinia* – see Chapter 6, and no. 63 below), both of which he revolutionized in the process.

18 **William John MacQuorn Rankine** (1820–72). He played a pivotal role in harnessing the emerging laws of thermodynamics for practical use, most notably in the Rankine Cycle (see Chapter 6), which still accounts for 85 per cent of all electricity produced worldwide. He devised what amounts to the Fahrenheit equivalent of the Kelvin absolute temperature scale, and the resultant Rankine Scale is still used by engineers in the United States. But his fundamental contribution to thermodynamic theory is his most lasting monument, in that it reinterpreted the work of Carnot and Clapeyron, recasting the theory in terms still recognizable to physicists and thermofluids engineers today.

19 **Alessandro Giuseppe Antonio Anastasio Volta** (1745–1827). An experimental physicist who made several major discoveries, including the first identification of methane. He also undertook early electrochemical experiments. It is for his work on early capacitors, and his invention of the electric battery (or 'voltaic pile'), that he is best remembered. The unit of electrical potential, the volt, is named after him.

20 **James Watt** (1736–1819). While working on a model Newcomen steam engine, Watt realized that its efficiency could be greatly improved if the exhaust were passed through a separate condenser: an insight that remains crucial to the efficient operation of all thermal power plants to this day. He entered business with Matthew Boulton of Birmingham, and Boulton–Watt engines went on to become the predominant energy conversion technology of the Industrial Revolution. The SI unit of power – the watt (see Chapter 4) – is named in his honour.

Twenty world records for energy engineering

21 **Largest biomass-fired power station:** Drax, UK, 0.67 GW

22 **Largest coal-fired power station:** Taichung, Taiwan, 5.5 GW

23 **Largest gas-fired power station:** Surgut-2, Russia, 5.6 GW

24 **Largest geothermal power plant:** Hellisheiði, Iceland, 0.3 GW

25 **Largest hydroelectric power station:** Three Gorges, China, 22.5 GW, which is also the largest power station of any type in the world

26 **Largest nuclear power station:** Kashiwazaki-Kariwa, Japan, 8.2 GW

27 **Largest oil-fired power station:** Shoaiba, Saudi Arabia, 5.6 GW

28 **Largest pumped-storage hydropower plant:** Bath County, USA, 3.3 GW

29 **Largest tidal barrage power station:** Sihwa Lake, South Korea, 0.25 GW

30 **Largest solar PV power station:** Topaz, USA, 0.55 GW

31 **Largest solar thermal (CSP) plant:** Ivanpah, USA, 0.377 GW

32 **Largest wave power plant:** Aguçadoura, Portugal, 0.002 GW

33 **Largest wind farm (onshore):** Alta, USA, 1 GW

34 **Largest wind farm (offshore):** London Array, UK, 0.63 GW

35 **Oldest district-heating system still in use:** Denver, Colorado (USA)

36 **Largest district-heating system:** Moscow, Russia

37 **Earliest commercial oil well:** Williams No. 1 Well, Oil Springs, Ontario, drilled in 1858

38 **Earliest industrial-scale coal mining:** Whickham Fell, Gateshead, England; commenced final quarter of the sixteenth century when Queen Elizabeth I liberalized the coal trade

39 **Oldest renewable energy power station still in use:** Ames Hydroelectric Generating Plant, near Ophir, Colorado, built in 1891; currently rated at 3.75 MW

40 World's oldest offshore wind farm: Vindeby, Denmark, installed in 1991. Total capacity 4.95 MW, comprising eleven 450-kW turbines

Three online energy tools

41 Energy efficiency without compromising beauty: some of Europe's most beautiful cities comprise historic buildings that have very poor energy efficiency. There is a perceived tension between renovating such buildings to modern standards and maintaining heritage features. This problem has been the focus of a project in Edinburgh, Scotland, the central districts of which are a World Heritage Site. An interactive guide to practical measures to resolve this tension is available here: www.ewht.org.uk/looking-after-our-heritage/energy-efficiency-and-sustainability/energy-efficiency-historic-buildings

42 See what's powering your national grid right now: it is possible to get an instantaneous read-out of the power sources contributing to several national grid systems in Europe using powerful, free online tools. For example, here's the link for the UK: www.gridwatch.templar.co.uk; and here is one for Spain: https://demanda.ree.es/movil/peninsula/demanda/total

43 Assess alternative pathways to achieve 2050 energy decarbonization targets: the UK government's Department of Energy and Climate Change has produced an online model – with output displayed in easy-to-read graphs – that lets you test out alternative blends of energy technologies to achieve future targets for low-carbon energy. It's a good antidote to woolly, wishful thinking. Try it here: http://2050-calculator-tool.decc.gov.uk/pathways/

Twenty energy heritage sites to visit

Coal heritage

There are numerous places where the history of the coal industry can be appreciated; here are just a few of the best places to visit:

44 The Mining Institute (www.mininginstitute.org.uk), Newcastle upon Tyne, England. This veritable cathedral of mining engineering is not only beautiful to behold but also houses the world's premier library on mining and the use of coal.

45 The National Coal Mining Museum for England (www.ncm.org.uk), Caphouse Colliery, near Wakefield, Yorkshire. This has an excellent underground gallery displaying mining technology right up to the most modern equipment, in authentic workings accessed by the oldest mineshaft still in use in the UK.

46 Arigna Mining Experience (www.arignaminingexperience.ie), near Carrick-on-Shannon, Republic of Ireland. A guided tour through a genuine drift mine that formerly worked both coal and iron.

47 Museo de la Minería de Asturias (www.mumi.es), El Entrego, Spain. An extremely impressive tour in a reconstructed gallery accessed by an original shaft, demonstrating the techniques used to work coal in the steep and faulted strata of northern Spain.

48 Zollverein Coal Mining Complex (www.zollverein.de), Essen, Germany. A World Heritage Site providing an excellent insight into the major coal-supported industries of the Ruhr Valley.

Oil and gas heritage

The oil and gas industry is still active worldwide and is not yet so much an object of nostalgia as former coal industries; neither are its visible remains as impressive. However, here are four museums that preserve early and more recent history of the industry:

49 The Oil Museum of Canada (www.lclmg.org/lclmg/Museums/OilMuseumofCanada/tabid/114/Default.aspx), Oil Springs, Ontario. Built on the site of the first successful oil well in the world, completed in 1858, this museum pipped at the post the following site (which also claims to be the oldest oil well in the world ...)

50 The Drake Well Museum (www.drakewell.org), Titusville, Pennsylvania. This museum is on the site of the first successful oil well in the USA, which entered production in 1859.

51 The Norwegian Petroleum Museum (www.norskolje.museum.no), Stavanger, Norway. Housed in a purpose-built building resembling an offshore oil platform, this is a truly exceptional museum telling the story of North Sea oil and gas from their first discovery in the 1960s.

52 The National Gas Museum (www.nationalgasmuseum.org.uk), Leicester, England. This is the largest collection of gas industry artefacts in the world, housed in the gatehouse of a former town gasworks.

Renewables heritage

53 New Lanark World Heritage Site (www.newlanark.org), central Scotland. This is a fully restored industrial model village, in which cotton mills were powered entirely by water from the adjoining river Clyde. Two working run-of-river hydropower stations operate to this day, and parts of the restored mill site are once again using water energy in the form of micro-hydro.

54 Geothermal Energy Exhibition (http://orkusyn.is), Hellisheiði, Iceland. At the world's largest geothermal power plant, this purpose-built visitor centre offers highly informative guided tours that let you see first-hand the operating turbines, as well as excellent scale models and films explaining the nature of geothermal energy and how it is harnessed.

55 Cruachan – The Hollow Mountain (www.visitcruachan. co.uk), by Loch Awe, Scotland. This is world's oldest pumped-storage hydropower station, still very much in operation. An excellent underground tour by minibus lets you see the vast turbine hall in the heart of the mountain. Although it does store grid electricity that is non-renewable, its systems of tunnels give it a far larger natural catchment than you would guess from a standard map, and the technology on display is that of classic, large-scale hydropower.

56 Centre for Alternative Technology (Canolfan y Dechnoleg Amgen) (www.cat.org.uk), Machynlleth, Wales. This is a living, working site which has already earned its place in the future heritage of renewable energy: it is the place that championed and showcased all the renewables long before most people wanted to know.

57 The Tide Mill Living Museum (www.woodbridgetidemill. org.uk), Suffolk, England. An amazingly enduring piece of renewable energy technology, this is a water mill operated by changing tidal heads. A mill has been operating on this site for over 840 years, and the present equipment lay dormant for just ten years from 1958 before being restored to full working order.

58 Kinderdijk World Heritage Centre (www.kinderdijk.com), west-central Netherlands. These 19 historic windmills

have been used since the eighteenth century to pump water out of the polders of the Randstad conurbation (which includes the four biggest cities in the country).

Other energy heritage sites

59 Energi Museet (www.energimuseet.dk), Bjerringbro, Denmark. This wide-ranging museum covers energy in nature, all major forms of energy production and use (fossil, nuclear and renewable), the environmental and climate change context, and even energy politics.

60 American Museum of Science and Energy (http://amse. org/), Oak Ridge, Tennessee, USA. Located near the Oak Ridge National Laboratory, which played a key role in the early development of nuclear power, the museum is, unsurprisingly, one of the best places to learn about nuclear energy (including its controversial ties to weaponry), but it also has striking interactive displays on many other energy resources.

61 *Turbinia* (www.twmuseums.org.uk/discovery.html). The world's first turbine-powered craft is housed in a purpose-built gallery at the Discovery Museum, Newcastle upon Tyne, England. This was the very vessel which Charles Parson cheekily used to make the Establishment sit up and pay attention to turbo-power!

62 Energie-Museum Berlin (www.energie-museum.de), based in a former battery storage facility attached to a disused substation. Run by volunteers and accessible only by prebooked guided tour, this is a large and impressive collection of equipment relating to power engineering, district heating and public lighting systems.

63 Energy Gallery at the Science Museum (www. sciencemuseum.org.uk/on-line/energy), South Kensington, London. Although aimed at children aged 7–14, this gallery is a learning experience for kids of all ages!

Half a dozen energy movies worth watching

The following two films include vivid depictions of the hardships which the mining of coal entailed a century or so ago:

64 *Germinal* (1993): historical drama directed by Claude Berri. Based on the classic novel by Emile Zola, it depicts the spiral of chaos as nineteenth-century coal miners in northern France strike for better conditions, and the mine owners respond violently.

65 *Matewan* (1987): historical drama written and directed by John Sayles. This depicts one of many largely forgotten instances of violent suppression by hired assassins of attempts to unionize in the Appalachian coalfields.

The only mainstream movie I am aware of that accurately depicts early oil industry activities:

66 *There Will Be Blood* (2007): historical drama written and directed by Paul Thomas Anderson, based on the novel *Oil* by Upton Sinclair. Set in early twentieth-century California, it charts the moral decline of a successful oil prospector in inverse proportion to his prosperity.

The first Hollywood treatment of the fracking revolution in the USA:

67 *Promised Land* (2012): drama directed by Gus Van Sant. This concerns attempts by a fictional company to persuade small-town landowners to sign up for shale gas drilling and fracking on their land. The gist of the story relates to corporate politics and a bungled attempt to pre-empt possible environmental objections by an agent provocateur.

68 *Edison the Man* (1940): drama directed by Clarence Brown. This is a hagiographical reworking of the life of Thomas Edison (see no.11 above), in the form of reminiscences by the aged gentleman on his lifetime achievements. Although most are fairly attributed, some of the claimed achievements in electricity were actually the work of others, such as Nikola Tesla.

Nuclear is an awkward one for Hollywood, with most movies (e.g. *Chernobyl Diaries*) being ludicrous mutant mayhem fantasies. Here's a credible one, based on a real case that happened in Oklahoma:

69 *Silkwood* (1983): drama/thriller directed by Mike Nichols. This is the story of a technician at a factory preparing plutonium fuel rods for nuclear power stations who becomes concerned that safety shortcuts are being taken as the plant falls behind on an important contract. When she raises the alarm, a chain of events ensues in which she is dangerously irradiated, psychologically manipulated and eventually killed in suspicious circumstances.

Four DIY energy projects

Do try these at home ...

70 Insulate to accumulate: the first thing anyone should do as a DIY energy project is minimize pointless losses of heat (or, if you live in a hot climate, loss of expensive chilled air) from your home. A good starting point is the Energy Savings Trust: www.energysavingtrust.org.uk/Insulation

71 Passive solar air heater made from aluminium cans: if you save up a pile of empty drink cans, you can use them to make a simple air heater – ideal for taking the chill out of a garden shed or summerhouse in the spring or autumn. http://www.ehow.co.uk/how_6679140_make-panel-using-aluminum-cans.html

72 A small wind turbine made from PVC pipe and wood: this device can be knocked together using tools commonly found in many domestic workshops. You need to buy some of the electrical components, but the superstructure is all made from low-cost materials. http://www.folkecenter.net/mediafiles/folkecenter/pdf/diy-wind-turbine.pdf

73 **Your very own pico-hydropower generator:** you could extract energy from the winter rains flowing down a drainpipe, or from the garden hose flow in the summer, using this simple hydropower turbine. http://www.greenoptimistic.com/2010/03/09/build-small-scale-hydroelectric-generator/#.UzBDVPl_sjw

Energy footprints: energy embodied in eighteen common commodities

Most people are now familiar with the concept of a 'carbon footprint' and will happily bandy around estimates of how much CO_2 has been released to produce, say, a tonne of aluminium. There's a big problem with this, however: carbon footprints are derived from the more fundamental energy footprints, which are far less commonly discussed. But the crux of the matter is that the same amount of energy will be used to produce a tonne of aluminium from identical ore in Iceland as in India – yet in Iceland the energy used in the aluminium smelter is from hydropower and geothermal sources, and is thus very low-carbon, whereas in India the smelters are coal-fired. As we attempt to decarbonize industry, we run the risk of failing to give credit where it is due if we casually talk about carbon footprints rather than energy footprints, from which we can then calculate the carbon footprint given accurate information on the industry in question. To encourage this more scientific approach, here are energy footprints for 20 common commodities. The values have been obtained from the University of Bath's excellent Inventory of Carbon and Energy (ICE) Database, which is freely accessible online via the following link: www.circularecology.com/ice-database.html. Each value given below is an answer (in units of MJ/kg, except where otherwise stated) to the question: 'How much energy does it take to make X?'

74 Aluminium 218 (primary); 29 (recycled)

75 Bricks 3 (amounting to about 7 MJ per brick)

76 Carpet 74 (i.e. about 187 MJ per square metre)

77 Cement 4.5

78 Ceramics 10

79 Clay roof tiles 6.5

80 Concrete 0.75

81 Copper tubing 57 (primary); 16.5 (recycled)

82 Paint 70 (for two coats, 21 MJ per square metre)

83 Paper 28 (for one pack of A4 printer paper: about 70 MJ)

84 Plastics 80.5

85 Photovoltaic cells (polycrystalline silicon) 4,070 MJ per square metre

86 Road surfaces (asphalt) 2,509 MJ per square metre

87 Sealants and adhesives (e.g. epoxy resin) 137

88 Steel 20.1 (average); 35.4 (primary); 9.4 (recycled)

89 Stainless steel 56.7

90 Timber 10 (hardboard 16; MDF 11; plywood 15)

91 uPVC windows (double glazed) 2,300 MJ per window

Nine organizations to check out if you care about some of the issues raised in this book

The following are listed here because I have personally found them to be trustworthy and objective as sources of information, and sincere in their dealings. This does not mean that I endorse everything they, or their parent organizations/sponsors, might say or do ...

92 **International Energy Agency** (www.iea.org): essentially the creature of the world's wealthiest countries (its member states), and to be consulted with that fact firmly in mind, it is a useful international source of statistics and information on trends in energy use and markets.

93 **United States Energy Information Administration** (www.eia.gov): a service of the US federal government that collates energy statistics from worldwide sources and produces useful summaries and analyses of trends.

94 **Annual Statistical Review of World Energy**, produced by the global petroleum corporation BP (http://tinyurl.com/kwyl98h): as this series of reports has been produced annually since 1951, by a consistent (if evolving) procedure, it gives a valuable insight into how a large private-sector organization views the world energy situation.

95 **The Energy Industry Times** (www.teitimes.com): the place to look if you want to keep up to date on the rapidly changing energy sector worldwide: a monthly newspaper, packed full of useful, objective features and insightful analysis.

96 **National Energy Action** (www.nea.org.uk): though covering only England, Wales and Northern Ireland, this UK charity gives an excellent insight into the issues of combating fuel poverty within a wealthy country.

97 WWF (http://wwf.panda.org): founded in 1961, this is now a world-leading NGO (non-governmental organization) that seeks to 'build a future in which humans live in harmony with nature'. WWF stands out from other major environmental NGOs in that it is refreshingly willing to listen to the sincere views of others, and, crucially, to alter its position on key issues if the scientific evidence warrants it. Check out the 'Energy 2050' report: http://tinyurl.com/ngtzn96

98 World Bank (www.worldbank.org/en/topic/energy): for all its faults, if the World Bank didn't exist we'd no doubt swiftly reinvent it. This site provides valuable perspectives from the developing world, providing a refreshing counterpoint to the IEA, US and BP worldviews.

99 World Energy Council (www.worldenergy.org): the United Nations–accredited network with a focus on energy strategies and policies.

100 The Low-Carbon Energy Development Network (http://lcedn.com/): a UK-based network of researchers, policy-makers and practitioners seeking to find ways to assist communities in developing countries of the Global South to bypass damaging, high-carbon pathways to a socially just, prosperous future.

Sources and further reading

Andrews, J., and Jelley, N., *Energy Science: principles, technologies and impacts* (Oxford: Oxford University Press, 2007).

Banfield, M., 'Darwinism, doxology, and energy physics: the new sciences, the poetry and the poetics of Gerard Manley Hopkins', *Victorian Poetry* 45 (2007): 175–94.

Banks, D., *An Introduction to Thermogeology: ground source heating and cooling*, 2nd edn (Chichester: Wiley, 2012).

Boyle, G. (ed.), *Renewable Energy: power for a sustainable future*, 3rd edn (Oxford: Open University and Oxford University Press, 2012).

Cox., B., and Forshaw, J., *Why Does E = mc²? (And why should we care?)* (Boston, MA: Da Capo Press, 2009).

Devereux, C., Denny, M., and Whittingham, M.J., 'Minimal effects of wind turbines on the distribution of wintering farmland birds', *Journal of Applied Ecology* 45 (2008): 1689–94.

Dinçer, I., and Rosen, M.A., *Thermal Energy Storage: Systems and Applications*, 2nd edn (Hoboken: John Wiley & Sons Ltd, 2011).

Electricity Networks Strategy Group, *A Smart Grid Vision* (London: ENSG, 2009). Available online: http://tinyurl.com/czwzmxq (last accessed 12 March 2014).

Eriksson, O., 'Environmental technology assessment of natural gas compared to biogas', Chapter 6 in P. Potocnik (ed.), *Natural Gas* (Rijeka, Croatia: INTECH/Sciyo, 2010).

European CCS Demonstration Project Network, 'Lessons learned from the Jänschwalde project: Summary report', Cottbuss, 2012. Available online: http://tinyurl.com/mlypq63.

Everett, B., Boyle, G., Peake, S., and Ramage, J. (eds), *Energy Systems and Sustainability: power for a sustainable future*, 2nd edn (Oxford: Open University and Oxford University Press, 2012).

Fraenkel, P., 'Underwater windmills: harnessing the world's marine currents', *Ingenia* 46 (2011): 13–18.

Freris, L., and Infield, D., *Renewable Energy in Power Systems* (Chichester: Wiley, 2008).

Gold, B.J., *ThermoPoetics: energy in Victorian literature and science* (Cambridge, MA: MIT Press, 2010).

Haszeldine, R.S., 'Carbon capture and storage: how green can black be?' *Science* 325 (2009): 1647–52. (DOI: 10.1126/science.1172246)

Hiemstra-van der Horst, G., and Hovorka, A.J., 'Fuelwood: the "other" renewable energy source for Africa?', *Biomass and Bioenergy* 33 (2009): 1605–16.

Intergovernmental Panel on Climate Change (IPCC), *Special Report on Renewable Energy Sources and Climate Change Mitigation.* Available online: http://srren.ipcc-wg3.de/report

Layton, B.E., 'A comparison of energy densities of prevalent energy sources in units of joules per cubic meter', *International Journal of Green Energy* 5 (2008): 438–55.

MacKay, D.J.C., *Sustainable Energy without the Hot Air* (Cambridge: UIT Publishers, 2008).

Martin, R., *SuperFuel: thorium, the green energy source for the future* (Basingstoke: Palgrave Macmillan, 2013).

McInnes, C., 'No time to abandon energy density', *Ingenia* 49 (2011): 12–13.

Nicola, S., and Andresen, T., 'Merkel's green shift forces Germany to burn more coal', *Bloomberg News*, 2012. Available online: http://tinyurl.com/8cdt8nc.

Roddy, D.J., 'Biofuels: environmental friend or foe?' *Proceedings of the Institution of Civil Engineers: Energy* 162 (2009): 121–30.

Roddy, D.J. (ed.), Biomass and Biofuels, *Comprehensive Renewable Energy* 5 (Amsterdam: Elsevier, 2012).

Roddy, D.J., and Younger, P.L., 'Underground coal gasification with CCS: a pathway to decarbonising industry', *Energy & Environmental Science* 3 (2010): 400–407 (doi:10.1039/b92119g).

Royal Academy of Engineering, *Future Ship Powering Options: exploring alternative methods of ship propulsion* (London: Royal Academy of Engineering, 2013). Available online: www.raeng.org.uk/futureshipping

Stephenson, M., *Returning Carbon to Nature: coal, carbon capture and storage* (Amsterdam: Elsevier, 2013).

Wade, N.S., Taylor, P.C., Lang, P.D., and Jones, P.R., 'Evaluating the benefits of an electrical energy storage system in a future smart grid', *Energy Policy* 38 (2010): 7180–88. http://dx.doi.org/10.1016/j.enpol.2010.07.045.

Westaway, R., and Younger, P.L., 'Accounting for palaeoclimate and topography: a rigorous approach to correction of the British geothermal dataset', *Geothermics* 48 (2013): 31–51. (doi:10.1016/j.geothermics.2013.03.009)

Wrangham, R., *Catching Fire: how cooking made us human* (New York: Basic Books, 2009).

Young, W., and Falkiner, R.H., 'Some design and construction features of the Cruachan Pumped Storage Project', *Proceedings of the Institution of Civil Engineers* 35/3 (1966): 407–50.

Younger, P.L., *Water: All That Matters* (London: Hodder & Stoughton, 2010).

Younger, P.L., Gluyas, J.G., Cox, M., and Roddy, D.J., 'Underground coal gasification', *Ingenia* 43 (2010): 42–6.

Index

accessibility, 32–3
alternating current (AC), 105–7
Ampère, André-Marie, 125
anaerobic digestion, 60–1
availability of energy, reduced, 116–18

baseload power, 50–1
batteries, 79
Becquerel, Antoine Henri, 126
Benz, Karl, 74
biodiesel, 75
biomass, 58–62
Brayton Cycle, 83–4
Buddle, John, 126

carbon capture and storage (CCS),
    37–8, 103–4
carbon emissions, 36–7, 102–4
carbon footprints, 139–40
Carnot, Sadi, 126
cars, electrically powered, 78–9
Celsius, Anders, 126
Churchill, Winston, 99
Clapeyron, Benoît Paul Emile, 127
Clausius, Rudolf, 21, 126–7
climate change, 118–23
coal, 68–9, 73, 133
    preparation, 55–6
Combined Cycle Gas Turbines
    (CCGTs), 84, 103
combined heat and power (CHP)
    systems, 54
combustion engines, 74–5
concentrated solar power (CSP)
    plants, 95
conversion of energy, 12–14, 20–2,
    54–8
costs of energy, 120–4

Cox, Professor Brian, 16
Cruachan power station, 44–6
Curie, Marie, 127

Daimler, Gottlieb, 74
Dalton, John, 127
demand
    meeting, 41–52, 101–2
    minimization, 111
density of energy/power, 33–5
Diesel, Rudolf, 75, 127
direct current (DC), 105–7
dispatchable power, 50–1

$E = mc^2$, 12–19
Edison, Thomas, 128
Einstein, Albert, 12, 17–19, 128
electrical charge storage, 44–7
electricity, 80–98
    distribution, 104–11
    in transport, 77–9
electrolysis, 76–7
energy saving projects, 138–9
energy security, 116–18
energy system, 40–2
enthalpy, 22
entropy, 21
evolution, human, 5

Faraday, Michael, 128
Feynman, Richard P., 13
films on energy, 137–8
food energy, 4–5
force, units of, 14–15
forms of energy, 12–14, 21
Forshaw, Jeff, 16
fossil fuels, 36–8
fracking, 56–7

fuel cells, 77–8, 79
fuel poverty, 123

gas turbines, 83–4
geothermal energy, 29, 32, 65, 71
geothermal wells, 55
Germany, 102–4, 117
gravity, 24–5, 27–8

Hadfield, Marie, 7
heat pumps, 69–71
Holliday, Steve, 102
Hopkins, Gerard Manley, 6–7
horsepower, 72–3
humans
  energy, 4–5
  muscle power, 72
hydrocarbons, 56–8
hydroelectric power, 84–6
hydroelectric pumped storage, 44–7
hydrogen, 76–7

international interconnector cables, 90

jet engines, 81–2
Joule, James Prescott, 128–9
joules, 14, 16–18

Kalina Cycle, 83
Kelvin, Lord, 7
King, Martin Luther Jr, 113

latent heat, 43–4
light, speed of, 16–17
liquefied natural gas (LNG), 75, 77–8

mass, 14–16
Merz, Charles Hesterman, 129

National Grid, UK, 48
national grids, 100–11, 132

natural gas, 56–8, 69, 134
newtons, 14–15
nuclear energy, 19, 30–1, 62–4, 84
  in transport, 75–6
nuclear fusion, 63–4

Ohm, Georg Simon, 129
oil, 134
oil refinery, 56–7
organic energy, 5–6
OYSTER device, 94

Parsons, Charles, 81, 129
Pelamis device, 94–5
phase-changing materials, 43–4
photovoltaic (PV) power, 95–8
poverty, 123
power density, 33–5
privatization of energy firms, 120
Prometheus, 23–4
protests, 114–15, 119
Pseudo-Hyginus, 23

radioactivity, 29–31
Rankine, William John MacQuorn, 129
Rankine Cycle, 82–4, 95
renewable energy, 64–5, 134–6
resources, 31–5

sewage, anaerobic digestion, 61
smart grids, 108–11
solar energy, 24–7, 29, 34, 95–8
solar heating, 71
Sommerfield, Arnold, 11
sources of energy, 24–31
specific energy, 33
steam locomotives, 73–4
Stephenson, George, 73
storage, 42–7
  electricity, 110
supply/demand, 41–52, 101–2

thermal energy, 43–4, 55, 68–71, 84
thermodynamics, 6–7
  laws, 19–22
thorium, 63
tidal energy, 27–8, 91–3
torrefaction, 59
transport, 72–9
Trevithick, Richard, 73
turbines, 80–98

units of energy, 13–14, 16–18, 49–50
uranium, 30–1, 62–3

vocabulary of energy, 6
Volta, Alessandro Giuseppe Antonio
    Anastasio, 130

waste energy, 54
water
  electrolysis, 76–7
  turbines, 82–3
Watt, James, 130
watts, 49–50
wave energy, 32, 93–5
weight, 14–15
Wells Turbine, 94
Wilde, Oscar, 39
wind energy, 76
wind turbines, 86–91, 109
wood, 68–9, 73
woody biomass, 58–9
World Heritage, 132
Wrangham, Richard, 5

ALL THAT MATTERS: ENERGY

All That Matters books are written by the world's leading experts, to introduce the most exciting and relevant areas of an important topic to students and general readers.

From *Bioethics* to *Muhammad* and *Philosophy* to *Sustainability*, the *All That Matters* series covers the most controversial and engaging topics from science, philosophy, history, religion and other fields. The authors are world-class academics or top public intellectuals, on a mission to bring the most interesting and challenging areas of their subject to new readers.

Each book contains a unique '100 Ideas' section, giving inspiration to readers whose interest has been piqued and who want to explore the subject further.